Simulation Method of Multipactor and Its Application in Satellite Microwave Components

Space Science, Technology and Application Series

Series Editors:
Weimin Bao, Yulin Deng

Simulation Method of Multipactor and Its Application in Satellite
Microwave Components
Wanzhao Cui, Yun Li, Hongtai Zhang, Jing Yang

For more information about this series, please visit: https://www.routledge.com/Space-Science-Technology-and-Application-Series/book-series/SSTA

Simulation Method of Multipactor and Its Application in Satellite Microwave Components

Wanzhao Cui
Yun Li
Hongtai Zhang
Jing Yang

CRC Press
Taylor & Francis Group
Boca Raton London New York

CRC Press is an imprint of the
Taylor & Francis Group, an **informa** business

北京理工大学出版社
BEIJING INSTITUTE OF TECHNOLOGY PRESS

First edition published 2022
by CRC Press
6000 Broken Sound Parkway NW, Suite 300, Boca Raton, FL 33487-2742

and by CRC Press
2 Park Square, Milton Park, Abingdon, Oxon, OX14 4RN

ISBN: 978-1-032-03897-1 (hbk)
ISBN: 978-1-032-03930-5 (pbk)
ISBN: 978-1-003-18979-4 (ebk)

DOI: 10.1201/9781003189794

Typeset in Minion
by codeMantra

Contents

Foreword I

FROM THE SUCCESSFUL LAUNCH AND OPERATION OF CHINA'S FIRST satellite Dong Fang Hong-1, to the successful launch and docking of Shen Zhou-9 and Tian Gong-1, and then to the successful networking and commercial application of Bei Dou satellite, China has become one of the world's largest aerospace powers, striding toward being a powerful aerospace country. With the continuous development of communications satellites, remote sensing satellites and navigation satellites, the demand of higher bit rate and more communication channels makes the launch power of spacecraft working in the microwave frequency range higher and higher, which inevitably brings a special physical effect under microwave high power, namely multipactor. Multipactor refers to a basic physical phenomenon that occurs in the vacuum environment. Under the condition of a certain radio frequency electromagnetic field, the electron will accelerate to move after getting enough energy, and then, it will trigger the secondary electron emission after colliding with the device. The rapid multiplication of such secondary electrons will lead to discharge breakdown on the surface of the microwave components. It affects the signal transmission integrity and electrical property of the whole system. It can even lead to gas discharge, causing permanent damage of microwave components, interruption of communication and system failure, resulting in spacecraft failure in orbit. The micro discharge effect is like the "cancer cell" of the spacecraft system, and the related key issues have a great impact on spacecraft.

As far as I know, in the 1980s, National Key Laboratory of Science and Technology on Space Microwave in China Academy of Space Technology (Xi'An) began to study multipactor. Later, they developed a device to measure multipactor and formed a systematic analytic theory and simulation method. The authors' research on multipactor is systematic and in-depth.

This book reflects their work on the numerical simulation method of multipactor. It is of important reference value for guiding multipactor engineering practice and basic scientific research in related fields. It is suitable for graduate students and technicians engaged in research and application in this field for reference.

It is beneficial to open the book. I sincerely hope that every reader can have a comprehensive understanding of the "multipactor effect," and it would be better if they can have a strong interest in it.

Yue Hao
Academician of CAS

Foreword II

THE MULTIPACTOR EFFECT HAS TROUBLED AEROSPACE ENGINEERS FOR a long time. The elimination of multipactor and its potential risk is the key link of high-power design. During my 24 years of designing microwave filters and passive components for communication systems as chief scientist and R&D director of COM DEV company in Canada, I encountered multipactor effect many times. I published a thesis on high-power filter design in *IEEE Microwave Magazine* [1], which summarized in detail the challenges in various high-power designs.

For engineering designers, the multipactor effect is a physical problem which involves material surface characteristics, vacuum electronics, electromagnetic field and particle interaction and other fields; for natural science researchers, the multipactor effect (also known as secondary electron emission) is an engineering phenomenon, which causes power loss, signal deterioration and transmission decline of the whole microwave transmission system. In engineering design, engineers often propose specific indicators to evaluate the anti-multipactor performance of devices. Because of its particularity and interdisciplinary, the multipactor effect has become a research hotspot in many fields.

The analysis of the multipactor effect is necessary to eliminate the risk. In the middle of last century, engineers proposed the equivalent model analysis method of multipactor effect breakdown voltage based on parallel plate approximation and gave the "sensitive curve" as the basis of engineering design. All along, this method has been used as a design standard for aerospace engineers in Europe and the United States. However, with the rapid development of microwave technology, the design of new filters, multiplexers, especially high-power microwave components presents fresh challenges. Since the 1990s, with the emergence of computational electromagnetics, it is possible to simulate the multipactor effect by a numerical method, which gradually became a trend. A new chapter of numerical simulation of the multipactor effect had begun.

At present, the research on the multipactor effect is mainly carried out from the aspects of material characteristics, calculation methods and suppression technology. There have been some theoretical research results formed by universities and research institutions at home and abroad, but no book on numerical simulation of multipactor effect has been published. Numerical simulation technology of the multipactor effect is interdisciplinary, which requires researchers to have various professional backgrounds; on the other hand, the development of modern electromagnetics opens up a new way. Under such a background, the timely book of *Simulation Method of Multipactor and Its Application in Satellite Microwave Components* is an inevitable result of the development of professional technology and the refinement of discipline classification, which fills the blank of related basic research books in China.

Based on the latest technical achievements, the book introduces the secondary electron emission characteristics of materials, comprehensively describes how to simulate the complete physical process of the multipactor effect based on the numerical model and shows the simulation process and results of examples. The profundity of this book is achieved with breathtaking lightness, and it is a great reference book for both researchers in basic science and design engineers of communication systems. For satellite system designers or other readers, it is also an excellent introduction to the high-power effect in space.

Professor Ming Yu
Department of Electronic Engineering, Chinese University of Hong Kong
IEEE fellow, member of Canadian Academy of Engineering[1]

---

[1] Ming Yu, "Power Handling Capabilities for RF Filters," *IEEE Microwave Magazine*, Oct. 2007, pp. 88–97.

Foreword III

IN THE HEART OF EVERY CHINESE, THERE IS A BEAUTIFUL MYTH ABOUT Chang'e and the rabbit. Exploring the vast universe has been a wonderful dream of mankind for thousands of years. Through the ages, Chinese people have never stopped exploring the universe. From the first man-made satellite carrying "The East is Red" resounding through space in 1970, and now Chang'e-3 is fully loaded with our dream to go into space, China's space industry has opened up a new chapter. During the six years from 1999 to 2005, from Shenzhou-1 to Shenzhou-6, six spaceships were launched six times. The speed and efficiency of China's manned spaceflight were amazing to the world.

With the development of China's space technology, a series of technical problems will be encountered in the implementation of aerospace projects. Multipactor breakdown is one of them. In extreme cases, multipactor breakdown can cause permanent damage to the microwave communication system on the aircraft, which cannot be repaired or replaced in orbit. As frontline researchers of aerospace engineering, the authors write this book with their strong theoretical foundation and vast engineering experience. This book describes the physical mechanism of multipactor breakdown and secondary electron emission, introduces multipactor simulation of microwave components and its application in travelling wave tube, and summarizes all problems and solutions of multipactor breakdown. It is a rare scientific and technological book closely related to engineering.

In my reading, there is no boring theoretical explanation in this book, it introduces the problem of multipactor breakdown from the generation mechanism and describes the simulation method in engineering application, which is very useful, practical and operable. I am not an expert in the multipactor breakdown, but I have a more comprehensive understanding of this complex problem only after reading some chapters. I really admire the frontline researchers for their strong theoretical foundation and rich engineering experience.

I have been in space all my life. In my opinion, to build China into a space power, we must strengthen the basic theoretical research like this. Only by understanding the problem of multipactor breakdown can we find out a set of effective solutions and establish a perfect theory of multipactor breakdown. I am well aware of the time and effort it takes to write such a book. The authors of this book are all frontline researchers of aerospace engineering. How hard it was for them to complete this book in addition to heavy scientific research and aerospace missions! I believe that while reading this book, you will not only gain the technical knowledge about multipactor breakdown, but also feel the enthusiasm of the authors sharing their theoretical knowledge and practical experience, and feel the aerospace spirit of hard work, courage to tackle difficulties, and innovation.

I really admire the older aerospace researchers' achievements under arduous conditions. It is gratifying to see that new aerospace researchers have made great efforts to overcome technical problems. I hope that more researchers are willing to share their theoretical knowledge and practical experience like the authors, so that engineers, teachers and students of relevant majors can get help to boost the follow-up development of China's space industry and space technology.

Shizhong Yang
Academician of Chinese Academy of Engineering

Preface

WITH THE DEVELOPMENT OF SPACECRAFT FOR COMMUNICATION, navigation and other services have changed and will further change the world's way of cooperation and people's lifestyle. For the satellite high-power microwave system, one of the most prominent problems is multipactor. Multipactor is a kind of physical phenomenon in which secondary electron emission occurs on the surface of material under specific conditions and synchronizes with the phase change of electromagnetic field in the RF vacuum tube, waveguide and other components, resulting in electron resonance multiplication, even an electron avalanche and current discharge. It may lead to surface damage and permanent damage of components. The numerical simulation of multipactor can fully analyse and verify the spacecraft on the orbit environment, reduce repeated design, avoid the long-term ground test, make it possible to intelligently design the high power and complex components in the spacecraft system and effectively improve the level and ability of spacecraft-advanced intelligent manufacturing.

This book summarizes the new methods, new technologies and new achievements of the numerical simulation technology of multipactor in recent years. Focusing on the spacecraft system and space application, the book comprehensively explains and expounds the three-dimensional numerical simulation technology of multipactor around the contents of theory, modelling, algorithm principle and simulation examples, focusing on the simulation and analysis software MSAT and its application. This book is for frontline researchers in the field of spacecraft and for relevant researchers in relevant universities, research institutes and college students. It can not only be used as a research reference for personnel in the professional field but also as a reference book for college students.

The book consists of six chapters, which were written and overseen by Wanzhao Cui, Yun Li, Hongtai Zhang and Jing Yang. Wanzhao Cui, Yun Li and Hongtai Zhang completed the unified draft of the book. Yun Li was responsible for the first and fourth chapters, Kai Peng was responsible for the third chapter, Jing Yang, Guobao Feng and Chunjiang Bai were responsible for the second and sixth chapters, Xinbo Wang completed the fifth chapter, Hongguang Wang, Yongdong Li and Jianfeng Zhang participated in the writing of the rest of the fourth chapter and Jianfeng Zhang participated in the revision of the third chapter.

Professor Yue Hao, academician of Chinese Academy of Sciences, and Professor Shizhong Yang, academician of Canadian Academy of Engineering and Professor Ming Yu, IEEE fellow, wrote the Forewords of this book and put forward many valuable opinions. The research work of this book is supported by key project of National Natural Science Foundation of China (No.: u1537211) and National Natural Science Foundation of China (No.: 11705142, 11605135, 61701394, 61801376, 51675421, 11675278, and 51827809). The publication of this book has been supported by the National Book Publishing Funding project, the 13th five year plan national key publication planning project and the national important equipment publishing project. The book has also been supported by Jun Li, vice president of China Academy of space technology (Xi'an), chief engineer Hongxi Yu, deputy director, Xiaojun Li, director Huang Puming of communication satellite business department, Professor Chunliang Liu of Xi'an Jiaotong University, and Professor Tiejun Cui of Southeast University. Tiancun Hu, Xinbo Wang, Na Zhang, Rui Wang, Qi Wang, Guanghui Miao, Guibai Xie, Yun He, Xiang Chen, He Bai and Huan Wei have carried out a lot of work for the study of multipactor. In the process of writing and editing, the book has been carefully reviewed by the editorial department of Beijing Institute of Technology Press, especially the guidance of editors Jialei Wang and Haili Zhang. Here, the author also expresses his thanks.

However, with the rapid development of science and technology, although the author has spent more than ten years in the process of writing the book, it has not fully covered all aspects of numerical simulation of multipactor. At the same time, with the development of supercomputing technology and modern electromagnetic computing, the numerical simulation technology of multipactor will make greater progress. Although every effort has been made to optimize the writing, but limited by the current research level and our capability, there are inevitably some omissions and deficiencies. We sincerely hope that the readers and experts will point our mistakes so that they will be corrected.

The Authors

This book combines the experience and achievements of engineering practice of China Academy of Space Technology, Xi'an, in the field of high-power multipactor in the past decades. It mainly introduces the related concepts, theories, methods and latest technologies of multipactor simulation, which are both theoretical and engineering. It also gives a comprehensive introduction to the outstanding progress made in the research technology of multipactor numerical simulation in China. At the same time, a three-dimensional numerical simulation method of multipactor for typical high-power microwave components of spacecraft is introduced, which is instructive and inspiring and has the corresponding technical depth.

This book is mainly for engineers in the field of high-power microwave technology. It can also be used as a reference for researchers in related fields, or as a teaching reference book for graduate students majoring in astronautics in colleges and universities.

CHAPTER 1

Introduction

1.1 OVERVIEW

Secondary electron multipacting effect, commonly known as multipactor, refers to a physical phenomenon that Secondary Electron Emission (SEE) occurs on the surface of material under certain conditions in the Radio Frequency (RF) vacuum tube, waveguide and other devices and synchronizes with the phase change of the time harmonic electromagnetic field, resulting in electron resonance multipacting, even electron avalanche and discharge. This effect may lead to surface damage or permanent damage of the device. In a vacuum environment, when electrons are accelerated by the RF electromagnetic field and collide with the material surface leading to SEE, then the yield of secondary electrons is related to the collision energy and angle between electrons and the material surface. One or more secondary electrons are emitted from the material surface. The mechanism of the multipactor effect is mainly divided into two categories, bilateral multipacting and unilateral multipacting.

Multipactor was first discovered in 1924 by the French physicist Camille Gutton, who observed a secondary electron multipacting effect during vacuum experiments, but failed to provide a clear definition and reasonable physical explanation. It was not until 1934 that Farnsworth discovered an electron avalanche multiplication under the action of RF frequency electromagnetic fields, designed and manufactured electron tubes to better study this phenomenon, distinguished multipactor from other general discharge phenomena for the first time and defined

DOI: 10.1201/9781003189794-1

as the secondary electron multipactor effect, namely multipactor [1–3]. Although Farnsworth realized that multipactor can produce certain hazards, he deemed it as an effective way to amplify signals. After several years of research, he finally used this phenomenon to amplify a very weak signal, which enabled the TV camera tube to capture outdoor scenes without natural light.

Until the mid-1940s, multipactor began to be used to describe the physical phenomenon of the secondary electron resonance multiplication and even the occurrence of a discharge rather than the electron tube itself. Over the years, researchers in this field have conducted in-depth research on multipactor, attempting to discover the underlying physical causes of multipactor, and the method to suppress multipactor and exploit multipactor in some specific situations.

The early research on multipactor mainly focused on devices such as vacuum electron tube and accelerator. With the development of space industry and spacecraft technology, research on multipactor for space application has been carried out gradually, and industry regulations were first formed in international space organizations such as the National Aeronautics and Space Administration (NASA) and the European Space Agency (ESA).

A spacecraft, operating on various orbits, can be exposed to a variety of factors of the space environment, among which the main factors that have a significant impact on space activities are solar electromagnetic radiation, the neutral atmospheric Earth ionosphere, the space plasma Earth magnetic field, space debris and micrometeoroids radiated by charged particles in space. These space factors, individually or collectively, will interact with spacecraft operating on-orbit, triggering various space environment effects on spacecraft materials and electronic components, which, in turn, affect the safe operation of the spacecraft.

For the new generation of satellite communication systems, the inevitable future trends, such as more users, more channels and higher bit rates, translate to higher power capacity processing capacity requirements. Satellite systems are located on different working orbits, and their natural radiation environment of the universe has a large number of charged particles carrying a certain amount of energy [4]. With the increasing input power, the electromagnetic field in the key components of the satellite communication system, such as output multiplexer, waveguide cavity filter, switch matrix and antenna feeder, may accelerate the movement of charged particles and synchronize them with the phase change of the electromagnetic field, and there is a very high risk of multipactor. Once it occurs, it will cause a discharge breakdown on the component surface and even a series of chain reactions such as gas discharge, causing permanent damage to the device, affecting the working performance of the whole satellite, and becoming an important technical bottleneck restricting the development of high-power space microwave components to higher power capacity [5–17]. Especially for on-orbit spacecraft, failures caused by multipactor effect are instantaneous and sudden and are often catastrophic hard.

In a vacuum environment, the mean free path of charged particles moving in an electromagnetic field is usually larger than the smallest spacing in the region of the maximum electromagnetic field strength in the microwave component. Therefore, when the charged particles collide with microwave components at a certain energy and angle, the SEE will occur. If the number of secondary electrons emitted in each collision is always greater than the number of electrons absorbed, and the secondary electrons can always keep up with the phase change of the electromagnetic field, and the electrons can constantly gain energy and acceleration, the electrons could grow exponentially and then may lead to the electron avalanche, the so-called multipactor, in which the number of electrons over time is exponentially increased, as shown in Figure 1.1.

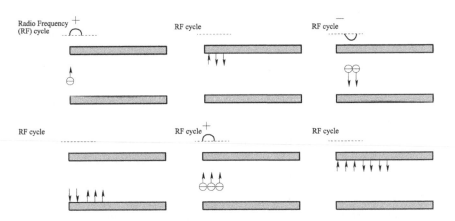

FIGURE 1.1   Schematic diagram of multipactor.

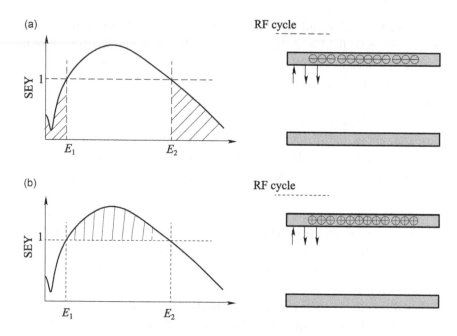

FIGURE 1.2 (a and b) Schematic diagram of the dielectric material surface charging principle.

For dielectric microwave components, considering the accumulation of charge on the surface of the dielectric, the multipactor process is much more complex than the metal multipactor. When the electron collides with the surface of the dielectric material, the surface of the dielectric material accumulates a large amount of negative charge, if the emitted electron is much less than the incident electron, as shown in Figure 1.2a. The dielectric material surface accumulates a large amount of positive charge after a period of time, if the emitted electron is much more than the incident electron, as shown in Figure 1.2b. Dielectric material surface charging affects both the physical process of SEE and the trajectory of the emitted electrons in space, so dielectric multipactor is a complex action process.

In summary, the following necessary conditions need to be met for the establishment of a single carrier multipactor effect under ideal conditions:

1.1.1 Vacuum Conditions

First, the basic condition for multipactor is that the mean free path of the electron is long enough to minimize the collision probability of the electron with the surrounding atom or molecule during the electron acceleration between two collisions in the device. When the air pressure is lower than 10^{-3} Pa, the mean free path is in the order of 10^{-1} m, which is comparable to the size of microwave components, and meets the necessary conditions for the occurrence of multipactor.

1.1.2 The Existence of Free Electrons

Second, multipactor must be excited by free electrons. There are many free electron sources in space, which can reach a large electron density, including solar wind and Van Allen belt. At an altitude of 750 km from the Earth, the density of free electrons in the ionosphere generally varies from 10^{-8} m^{-3}, depending on the solar activity cycle and the specific location of the spacecraft and providing a large source of free electrons for multipactor.

1.1.3 Maximum Secondary Electron Yield (SEY) of the Material Is Greater Than 1

The third basic condition for multipactor is that the maximum Secondary Electron Yield (SEY) of the material is greater than 1. As shown in Table 1.1, all of the maximum SEYs of commonly used typical aerospace materials are greater than 1. The SEY is related not only to the material but also to the incident energy of the primary electron, which depends on the power of the applied RF signal and the power obtained by the primary electron between collisions. Therefore, multipactor effect is difficult to establish, when the small power of the applied RF signal and the low incidence energy of the primary electron are unable to excite a larger number of secondary electrons. When the primary electron incidence energy is higher under the large external RF signal power, the primary electron can penetrate into the surface deeply, so that the secondary electrons produced cannot reach the surface due to the capture by the material internal molecules and/or atoms. Then, it is hard to excite enough SEE. Thus, the external RF signal power must be within a certain range to induce multipactor.

TABLE 1.1   Maximum SEY, δ_{max}, for Typical Materials

Material	Al_2O_3	BeO	CsI	Ag	Cu	Fe	W	Pt	Pd
δ_{max}	4	3.4	20	1.5	1.3	1.3	1.4	1.8	1.3

1.1.4 Transition Time of Secondary Electrons Is an Odd Multiple of One Half-Cycle of a Microwave Signal

Under the above-mentioned conditions for the occurrence of multipactor, multipactor is prone to occur in the high-power microwave components of the spacecraft, such as communication, navigation and remote sensing satellites. In addition to the above necessary conditions, the last necessary condition for the occurrence of multipactor is that the transit time of secondary electrons emitted from the material surface should be an odd multiple of one half period of the microwave signal, so that the secondary electrons emitted can always be accelerated under the action of the electromagnetic field and proceed to the next collision, so that the electronic resonance multiplication continues to occur. However, recent studies on unilateral multipactor and multi-carrier multipactor have shown that multipactor may occur within a sufficiently long single period under certain conditions. Therefore, this necessary condition is no longer valid under certain conditions.

According to the traditional analysis theory of multipactor, the multipactor threshold is closely related to the product of the operating frequency of the microwave signal and the gap size of the component ($f \times d$). The relationship between the multipactor threshold of the waveguide and $f \times d$ is shown in Figure 1.3. It can be seen that the smaller the $f \times d$, the lower the multipactor threshold. Especially in the low-frequency band, even if d is very large, the multipactor threshold is still very low, and multipactor will occur in microwave components under conditions of very small power.

For high-power microwave components operating in a vacuum or near-vacuum environment, multipactor is a surface breakdown effect due to the electron avalanche effect that is caused by the SEE of the metal or dielectric material under high-energy electron impact [8]. As the power level increases, the electrons moved in the electromagnetic field and collided with the microwave component surface with a certain momentum, generating SEE. If the electromagnetic field, such as frequency, field strength and phase, meets certain conditions, the secondary electrons emitted from a metal or dielectric material repeatedly travel between the surface of the metal or dielectric material, and a sustained multiplication of the electrons occurs, which will eventually trigger a discharge breakdown. Therefore, it appears as a resonant secondary electron multipacting effect, which is a non-linear process. When it occurs, a large amount of energy

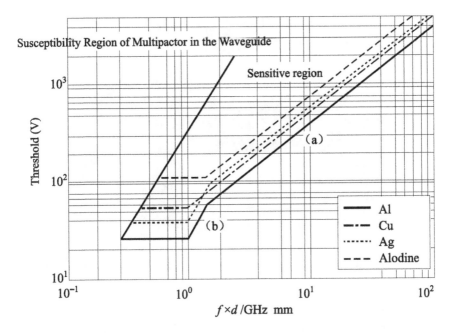

FIGURE 1.3 (a and b) Relation curve of multipactor threshold voltage with $f \times d$ of rectangular waveguide structure.

will be dissipated at a small point inside the component. Once multipactor occurs, it will cause serious consequences. Its main hazards include:

1. Detune the resonant component with high-quality factor, causing the maladjustment of the transmitted microwave power signal. The additional inductance effect caused by multipactor is a highly non-linear, random and time-varying short-term-induced impedance. This effect will cause fluctuations in the Q value, coupling parameters, losses and phase constants of the resonant cavity, which will inevitably lead to system detuning, resulting in performance degradation.

2. It leads to the escape of adsorbed gas and internal gas on the surface of the material, resulting in a more serious gas discharge. When multipactor occurs, the above-mentioned gas may be escaped. If the escaped gas cannot be discharged in an appropriate manner, low-pressure gas discharge may occur. The energy released by the gas discharge is more than that of multipactor, which is more likely to cause damage to the components, thereby causing the entire system to fail.

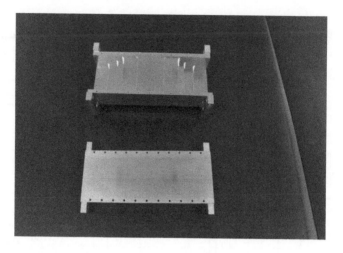

FIGURE 1.4   The damage of microwave components caused by multipactor.

3. Multipactor may cause the ablation of the cavity surface of the microwave component, destroy the surface state and conductivity, and deteriorate the performance of the component Figure 1.4.

4. Generate narrow-band noise near the carrier frequency.

5. Multipactor is an important non-linear factor in high-power components, which is one of the reasons for causing passive intermodulation.

1.2   RESEARCH BACKGROUND OF SPACECRAFT MULTIPACTOR EFFECT

Multipactor has become a hot topic in the space application. The satellite communications use microwave-frequency electromagnetic waves to communicate with base stations on the ground because low-frequency electromagnetic waves cannot pass through the ionosphere. Microwave frequencies also allow engineers to be more flexible in designing electronic devices. These small electronic devices can effectively reduce the payload weight and volume, which is very important for satellite systems. This also makes microwave systems working in vacuum more sensitive to multipactor. In space, multipactor breakdown is a critical issue because in extreme cases, it can cause permanent damage to the microwave communication system on the spacecraft, which is difficult or impossible to repair and replace on-orbit after launch.

With the development of modern satellite communication systems to multi-carrier and high-power communication, the prediction of multipactor breakdown threshold becomes increasingly important. In order to improve the power handling capability of RF microwave components, aerospace industry engineering researchers began to work on the simulation methods of the physical processes of multipactor breakdown in microwave structures. In order to achieve this goal, a more accurate model should be established to predict the breakdown level of various RF microwave components on the satellite, such as waveguide, filter, microwave switch and multiplexer.

The cosmic ray radiation generates a large number of free electrons on the equipment of the satellite. When these electrons exist in the electromagnetic field with enough electromagnetic strength that can cause the electrons to emit and leave the wall surface, the initial stage of multipactor is formed. Under vacuum, the electrons can be accelerated very quickly toward the other walls or surfaces because no gas particles collide with it and slow it down. When striking a wall, depending on the impact energy, incident angle and SEE characteristics of the wall surface, an impacted electron can release one or more secondary electrons into vacuum. If the RF signal changes phase synchronously, these secondary electrons can be accelerated toward the wall of another device and collide to produce more secondary electrons, which increases the electron density exponentially, resulting in multipactor. The establishment of multipactor can largely reflect the damage of incident power to the system.

Past work had focused on analytical methods for multipactor modelling and attempted to predict the multipactor breakdown voltage of simple geometries by obtaining analytical formulas for electron motion. In recent years, since most of the existing analytical models can only analyse specific physical structures and multipactor order and predict the multipactor breakdown voltage of geometry with limited structure due to their low calculation efficiency, it is important to develop new analysis methods and means.

1.3 RESEARCH HISTORY OF NUMERICAL SIMULATION METHODS OF MULTIPACTOR FOR SPACE APPLICATION

As the next-generation spacecraft load technology develops toward higher power, more channels, and smaller sizes, the mechanism, numerical simulation, suppression, and experimental research of the secondary electron multipactor effect (also known as multipactor effect) have gradually become the basic research hotspot of space microwave technology. The numerical simulation technology based on the physical mechanism of multipactor has also gained increasing attention as a key technical means for the design and suppression of multipactor in high-power microwave components.

Free electrons are generated when cosmic rays radiate a spacecraft collide with spacecraft microwave components under the action of the electromagnetic field to produce SEEs. If the electromagnetic driving force is strong enough that the average yield of the secondary electrons emitted is greater than the absorbed electrons, and the electron motion is consistent with the phase change of the electromagnetic field, multipactor will occur. Multipactor will cause a series of chain reactions such as surface damage, signal distortion, power loss and even gas discharge, which will permanently damage the device. It affects the performance of the whole satellite and becomes an important technical bottleneck restricting the development of high-power space microwave components. Particularly for spacecraft on-orbit, the failures caused by multipactor are instantaneous, sudden and are often catastrophic hard failures.

In combination with the phenomenological model of electrons and microwave components, the numerical simulation technology of multipactor usually analyses the trajectory of electrons driven by electromagnetic fields in vacuum and the laws of electron motion under different electromagnetic field distributions to reveal the change trend of the number of electrons with time under different electromagnetic field strengths and SEE characteristics of different components, to obtain physical images and threshold power of multipactor, which provides the necessary technical approach for the basic physics research of space plasma, and theoretical and technical guidance for the design of high-power microwave components of spacecraft.

Since the discovery and utilization of multipactor in the 1930s, the numerical analysis technology of multipactor has made great progress. On the basis of the accumulation of initial and development stages that have lasted almost a century, and with the further development of computer technology, several major space agencies in the world, including China, the National Space Administrations of the United States, Canada, Russia, Spain, Sweden, Japan and their subordinate universities and research institutions, have begun to carry out numerical analysis research studies on multipactor of high-power microwave components and have made some achievements.

The 1930s to the 1970s was the initial stage of multipactor analysis technology. During this period, multipactor analysis was mainly based on analytical approximate calculation of the electromagnetic field distribution of microwave components and the analytical solution of the multipactor electron motion dynamics to briefly explain the generation process of multipactor in microwave components. When the free electrons in vacuum electronic equipment leave the device surface and are accelerated to another equipment surface due to the high-power electromagnetic field, multipactor occurs. Depending on the strength of the electromagnetic field, these electrons with enough energy impact the wall to release more secondary electrons from the surface of the structure. If the direction of the field changes at the same time of impact, the emitted electrons can be accelerated to the opposite wall, impacting and releasing more electrons, which is called a resonance state. If this process continues for several RF cycles, the electron density increases exponentially and triggers a secondary electron multipactor effect discharge. Experimental results show that this exponential discharge will not continue to grow but will eventually reach saturation. The literature introduces some theories to explain such reasons, but most researchers believe that the most likely cause is the space charge effects. By equivalence of the electromagnetic field distribution under ideal conditions, the Lorentz equation that regulates the electron motion is solved to obtain the trajectory of the electron motion and the analytical solution of the numerical simulation of multipactor and then determining the input voltage threshold of multipactor. This is a common solution in the initial stage of the development of numerical multipactor simulation.

During this period, researchers have conducted many theoretical and experimental studies to record "multipactor Susceptibility Region" and the input power range that can initiate and maintain multipactor breakdown. For various gap sizes and frequency ranges, these graphs are used to scale the multipactor threshold by the product of frequency (f) and gap (d) ($f \times d$). These curves are used to design high-power RF devices that do not cause multipactor. Many research institutions and researchers have developed susceptibility curves suitable for different microwave components through theoretical and experimental methods.

J. Sombrin, an early researcher, published a research paper illustrating the boundaries between different multipactor orders [17]. Assuming that all electrons between the plates are emitted at a constant initial velocity V, i.e., theory of constant V, he pointed out that each multipactor order has its own susceptibility area, and these areas decrease with the increase of the multipactor order. However, Hatch and Williams found this assumption to be impractical (because the velocities of the secondary electrons are statistically random), and they developed similar curves to improve the constant initial velocity method [18].

Hatch and Williams proposed different ways to calculate the multipactor susceptibility curve of the parallel plate model because the method of constant initial velocity of electron emission in vacuum is not in line with physical reality. Hatch and Williams assumed that the ratio of the collision velocity of electrons to the material and the initial velocity of electron exit is constant k, instead of the constant initial velocity. In the phase resonance state, the boundary condition of the constant k theory is rewritten as [19]:

$$\alpha = \arctan\left[\frac{1}{k-1}\left(\frac{k\omega d}{v_1} - (k+1)n\frac{\pi}{2}\right)\right] \tag{1.1}$$

Compared to the multipactor susceptibility curve based on the assumption of constant velocity V theory, Hatch and Williams show multipactor regions with equal width, which is due to the constant ratio of the impact velocity of each electron to the initial velocity. Therefore, for a given multipactor order, a higher $f \times d$ product results in a larger bifurcation between the two graphs because the Sombrin region of the higher $f \times d$ product gradually becomes thinner. However, at the same multipactor order, the above researchers' results are in good agreement for the lower $f \times d$ product.

In the 1980s, Woode and Petit obtained multipactor results through a large number of experiments and used Hatch and Williams diagrams to curve-fit their experimental data [20]. To achieve this, Woode revised the constant k theory. Although this method is not in line with physical reality, the results in the multipactor susceptibility region chart are very consistent with the experimental data. From the perspective of the model, this process is not helpful for understanding multipactor. However, it is very useful from the perspective of engineering design. Therefore, the Woode and Petit curve has been widely adopted by the ESA spacecraft high-power microwave component design since it was proposed and has been promoted in the space industry as design standards.

Obviously, the impact of electron avalanches caused by multipactor cannot be sustained, and there must be some physical phenomenon to delay this process. In the late stage of multipactor from initialization, generation, formation to saturation, the growth of electron cloud volume is proportional to the increase of electron density (due to phase focusing). This will create a space charge effect (also known as the electron cloud effect), which acts on the electrons in the electron cloud through mutual repulsive forces between the electrons. This means that many electrons can enter the interior of the wall through their own field, so they hit the wall with very little energy and no secondary electrons are generated. This interaction can eventually reach an equilibrium state, at which the SEY is about $SEY = 1$. If the electron cloud becomes very large, eventually the interaction between the electrons is large enough to detune the electron cloud. As the electrons on the wall are increased through the SEE and positive charges can leave, the electrons leaving the wall surface are larger than the impacting ones. The positive charge will disturb the motion of secondary electrons leaving the surface, and the electrons will be detuned if the influence on the transmission time is large. The impact of these effects on the motion of electrons is mainly gained by empirical observation after a period of exponential increase in electron density. For deep breakdown, the rate of increase slows down and eventually stabilizes. The number of electrons that decrease as they leave the wall is equal to the number that increases as they collide. At this time, the multipactor susceptibility curve based on analytical solution can no longer explain the physical phenomenon.

From the 1990s till date, it is the development stage of multipactor analysis technology. During this period, multipactor analysis technology was mainly developed and gradually matured based on the latest achievements of interdisciplinary-related technology development, such as computer science, technology and surface science.

Since the 1990s, with the development of computer science and technology, especially the development of computational electromagnetics, new theoretical and technical approaches have been brought to the numerical simulation of multipactor. In order to successfully predict the onset of multipactor and record information about electron impact, such as the location, energy and time of the impact, the electron trajectory transiting the device must be calculated. Thus, the electron motion that varies with time needs to be determined, which can be achieved by solving the Lorentz force equation using numerical means. As a first-principles numerical calculation method, the particle-in-cell (PIC) technology is often used to track electron trajectories in plasma physics. Combining electromagnetic calculation theory and SEE modelling theory, the PIC technology can be used to track the evolution of electron motion during the initialization and formation of multipactor.

Major international aerospace institutions have supported universities and aerospace companies to carry out extensive numerical simulation studies of multipactor, which have established multipactor models of microwave components with typical structures such as slabs and rectangular waveguides, and developed Multipactor Calculator [21], FEST3D [22–26], MEST [27], MultP [28–30], TRAK-RF [31, 32] and other multipactor simulation software.

Through experiments, ESA has established multipactor susceptibility curves of microwave transmission lines such as micro-strip structures, rectangular waveguides, circular waveguides, and square coaxial lines. A large number of multipactor experiments have been conducted for different $f \times d$, and a multipactor simulation software, Multipactor Calculator, has been developed. The software is widely used by international RF engineers for the design of high-power components. Based on the electromagnetic field distribution obtained by the electromagnetic simulation software, the designer preliminarily determines the most susceptible area of multipactor, selects the integration path, obtains the voltage between the two plates, and compares this voltage with the multipactor threshold of the corresponding plate $f \times d$ in the Multipactor Calculator to determine the multipactor margin. The software can quickly obtain the multipactor threshold of microwave components, but it depends heavily on the designer's experience. Different region judgments and different integration path selections will cause the multipactor threshold to change, making the product less consistent. It is very effective for the waveguide components, but the prediction accuracy of the multipactor threshold for coaxial structures will decline, and it is no longer applicable to microwave components with more complex structures, such as coaxial cavity filters.

The analysis and research of multipactor of the traditional metal components are usually based on the infinite parallel plate approximation. In recent years, multipactor mechanism of closed structure microwave components, such as rectangular structure [33, 34], coaxial structure [11, 35], elliptical structure and ridge waveguide structure [36, 37], has been studied successively.

The PIC method based on first-principles, which simplifies the physical reality very little, is very suitable for simulating the non-linear interaction process between electrons and electromagnetic fields [38–52]. In the PIC method, macro particles or particle cloud models are used to represent the actual electrons in the space. It is with a volume of about the size of the Debye sphere. It is an advanced numerical simulation method to describe the interaction between microwave components and free electrons at present. It plays an important role in revealing the complex physical process and discovering new physical laws. With the support of ESA, Darmstadt University of Technology, Swiss Federal Institute of Advanced Technology and Tesat Company jointly carried out numerical simulation study on multipactor of microwave components based on PIC algorithm, and used the experimental results to modify the multipactor model. In 2006, a Co-developed FEST3D (Full-wave Electromagnetic Simulation Tool 3D), a numerical simulation software for multipactor, can only perform multipactor analysis for specific types of microwave components with rectangular waveguide structures, while ignoring the elastic and inelastic electron scattering processes in the collision between electrons and surface materials.

Richard et al. from Chalmers University of Technology in Sweden carried out numerical simulation study of multipactor on coaxial transmission lines. One-side and two-sides multipactor theories are established for coaxial lines with different inner and outer diameter ratios. Based on this theory, the 2D PIC simulation method of multipactor is carried out and used. It provides a theoretical basis for the multipactor analysis in microwave components with coaxial structure.

In order to solve the problem of multipactor analysis in microwave components with complex structures, M. A. Gusarova et al. of the Moscow Institute of Engineering Physics developed multipactor numerical simulation software MultP and its improved version MultP-M for typical structural accelerators, by which the process of electron motion is simulated based on the external electromagnetic field, without considering the interaction between electron and electromagnetic field. In combination with statistical distribution and the Monte Carlo method, the Chalmers University of Technology in Sweden, the French Space Agency and the ComDev Company in Canada, respectively, carried out multipactor

TABLE 1.2　Comparison of Typical Multipactor Simulation Software

Software	Multipactor Calculator	FEST3D	MEST	MultP	MSAT
Research Organization	ESA	University of Valencia, Spain; ESA	University of Madrid, Spain; ESA	Russia, INR and MEPI	China Academy of Space Technology, Xi'an, Southeast University, Xi'an Jiaotong University
Electromagnetic solution	Analytical solution	Typical structure electromagnetic calculation library	Analytical solution	MAFIA	FDTD method
Micro-discharge algorithm	Empirical data fitting curve	PIC	Monte Carlo	PIC	3D PIC

numerical simulations on specific microwave components using the PIC method. The performance comparison of some typical multipactor simulation software developed is shown in Table 1.2.

1.4 RELATED RESEARCH INSTITUTIONS AND RESEARCH PROGRESS IN CHINA

In the past ten years, some basic theoretical and numerical research studies work on multipactor have gradually been carried out by China Academy of Space Technology, Xi'an, Southeast University, Tsinghua University, Xi'an Jiaotong University, Zhejiang University, and the University of Electronic Science and Technology. Among them, the research team of China Academy of Space Technology, Xi'an, together with Xi'an Jiaotong University and Southeast University, after nearly 10 years of technology and research accumulation, systematically proposed a method for multipactor electromagnetic particle co-simulation and threshold analysis platform (Multipactor Simulation and Analysis Tool, MSAT). The software interface is shown in Figure 1.5.

The time-varying electromagnetic field distribution inside the microwave component is calculated by MSAT using the finite difference time domain (FDTD) method, and the time evolution process

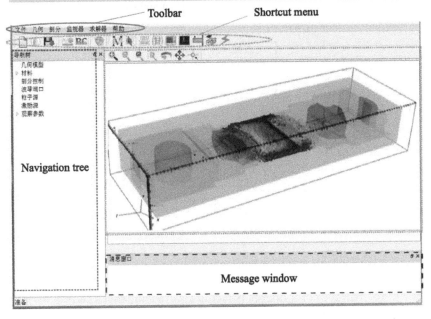

FIGURE 1.5   Software interface of multipactor simulation and analysis platform MSAT.

of the space electron inside the microwave component is calculated using PIC method. The principle of the electromagnetic particle co-simulation algorithm is shown in Figure 1.6. Through coupled electromagnetic field calculation and particle motion propulsion, using a SEE model that considers the actual working conditions of the surface of the metal microwave component, 3D multipactor simulation and threshold analysis are successfully achieved by MSAT, reproducing the complete physical process of multipactor initiation, evolution and saturation in three-dimensional space. According to the reported results, the actual multipactor threshold simulation results of microwave components are in good agreement with the multipactor test results.

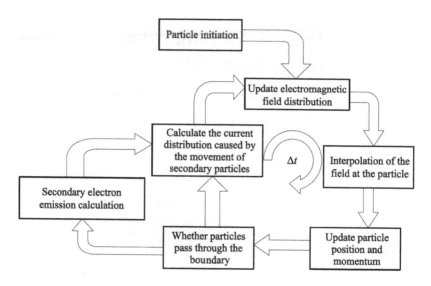

FIGURE 1.6 Multipactor simulation principle in MSAT.

1.5 SUMMARY

With the development of space science and technology, the reality of high-power microwave components is facing severe challenges caused by multipactor. This chapter reviews and summarizes the research results of the current multipactor numerical simulation methods in the main aerospace institutions and research institutes at home and abroad. On the basis of summarization and comparison, it points out the main research directions and latest progress of multipactor numerical simulation.

REFERENCES

1. Farnsworth PT. Television by Electron Image Scanning. *Journal of the Franklin Institute*. 1934, 2:411.
2. Gallagher WJ. The multipactor effect. *IEEE Transactions on Nuclear Science*. 1979, ns-26(3):4280–4282.
3. Paschke F. Note on the Mechanism of the Multipactor Effect. *Journal of Applied Physics*. 1961, 32:747–749.
4. Chu G, Ma S, Li T, Fan T. *Introduction to Aerospace Technology*. Beijing: Aviation Industry Press, 2002.
5. Woode A, Petit J. Investigation into multipactor breakdown in satellite microwave payloads. *ESA Journal*, 1990:467–468.
6. Rozario N, Lenzing H. Investigation of Telstar 4 spacecraft Ku-Band and C-Band antenna component for Multipactor breakdown. *IEEE Transactions on Microwave Theory and Techniques*, 1994, 42:558–564.

7. Kudsia C, Cameron R, Tang WC. Innovations in microwave filters and multiplexing networks for communications satellite systems. *IEEE Transactions on Microwave Theory and Techniques*, 1992, 40:1133–1149.

8. Vaughan JRM. Multipactor. *IEEE Transactions on Electron Devices*, 1988, ED-35:1172.

9. Hatch A, Williams H. The secondary electron resonance mechanism of low-pressure high-frequency gas breakdown. *Journal of Applied Physics*, 1954, 25:417–423.

10. Kim HC, Verboncoeur JP. Time-dependent physics of a single-surface multipactor discharge. Physics of Plasmas, 2005, 12:123504.

11. Semenov VE, Zharova N, Udiljak R. Multipactor in a coaxial transmission line. II. Particle-in-cell simulations. *Physics of Plasmas*, 2007, 14:033509.

12. Li Y, Cui WZ, Zhang N, Wang XB, Wang HG, Li YD and Zhang JF. Three-dimensional simulation method of multipactor in microwave components for high-power space application. *Chinese Physics B*, 2014, 23(4):048402.

13. Pivi M, King FK, Kirby RE, Raubenheimer TO, Stupakov G, Pimpec FL. Sharp reduction of the secondary electron emission yield from grooved surfaces. *Journal of Applied Physics*, 2008, 104:104904.

14. Ye M, He YN, Hu SG, Yang J, Wang R, Hu TC, Peng WB, Cui WZ. Investigation into anomalous total secondary electron yield for microporous Ag surface under oblique incidence conditions. *Journal of Applied Physics*, 2013, 114:10495.

15. Ye M, He YN, Hu SG, Wang R, Hu TC, Yang J, Cui WZ. Suppression of secondary electron yield by micro-porous array structure. *Journal of Applied Physics*, 2013, 113:074904.

16. Chang C, Liu GZ, Fang JY, Tang CX, Huang HJ, Chen CH, Zhang QY, Liang TZ, Zhu XX, Li JW. Field distribution, HPM multipactor, and plasma discharge on the periodic triangular surface. *Laser and Particle Beams*, 2010, 28(01):185.

17. Sombrin J. Effect multipactor. CNES Technical Report, 1983, 83.

18. Hatch AJ, Williams HB. Multipacting modes of high-frequency gaseous breakdown. *The Physical Review*, 1958, 112(3):681–685.

19. Udiljak R. Multipactor in low pressure gas and in nonuniform rf field structures. PhD Thesis-Chalmers University of Technology Sweden, 2007.

20. Woode A, Petit J. Diagnostic investigations into the multipactor effect and susceptibility zone measurements and parameters affecting a discharge. ESTEC Working Paper, 1989.

21. ESA-ESTEC. *Space Engineering: Multipacting Design and Test*. the Netherlands: ESA Publication Division, 2003, ECSS-20-01A.

22. Boria VE, Gimeno B. Waveguide filters for satellites. *IEEE Microwave Magazine*, 2007, 8(5):60–70.

23. Vicente C, Mattes M, Wolk D. FEST3D-a simulation tool for multipactor prediction, The 5th International Workshop on Multipactor, Corona and Passive Intermodulation in Space RF Hardware, Valencia, Spain, 2005.

24. Wolk D, Vicente C. An investigation of the effect of fringing fields on multipactor breakdown, The 5th International Workshop on Multipactor, Corona and Passive Intermodulation in Space RF Hardware, Valencia, Spain, pp. 12–15, 2005.

25. Anza S, Vicente C, Raboso D. Enhanced prediction of multipaction breakdown in passive waveguide components including space charge effects, IEEE MTT-S International Microwave Symposium Digest, pp. 1095–1098, Atlanta, GA, 2008.

26. Armendáriz J, Monge J, Ghilardi M. FEST3D simulation tool: recent advances and developments, The 6th International Workshop on Multipactor, Corona and Passive Intermodulation in Space RF Hardware, Valencia, Spain, 2008.

27. Lara J, Perez F, Alfonseca M, Garcia-Baquero DR. Multipactor prediction for on-board spacecraft rf equipment with the MEST software tool. *IEEE Transactions on Plasma Science*, 2006, 34(2):476–484.

28. Gusarova M, Kaminskii V. Evolution of 3d simulation multipactoring code MultP. *Problems of Atomic Science and Technology*, 2008:123–126.

29. Krawczyk FL. Status of multipacting simulation capabilities for SCRF applications. *Office of Scientific & Technical Information Technical Reports*, 2001.

30. Kravchuk L, Romanov G. Multipactoring code for 3d accelerating structures, The 20th International LINAC Conference, Monterey, CA, 2000.

31. Humphries S, Dionne N. Modeling secondary emission in a finite element multipactor code, Proceedings of the 9th Workshop on RF Superconductivity, 390–396, Santa Fe, New Mexico, 1999.

32. Humphries S, Rees D. Electron multipactor code for high power rf devices, electron multipactor code for high-power rf devices, Proceedings of Particle Accelerator Conference, 1997:2428–2430.

33. de Lara J, Perez F, Alfonseca M, Garcia-Baquero DR. Multipactor prediction for on-board spacecraft RF equipment with the MEST software tool. *IEEE Transactions on Plasma Science*, 2006, 34:476–484.

34. Vicente C, Mattes M, Wolk D, Mottet B, Hartnagel HL, Mosig JR. Multipactor breakdown prediction in rectangular waveguide based components, IEEE MTT-S International Microwave Symposium Digest, 1055–1058, California, 2005.

35. Udiljak R, Anderson D, Lisak M. Multipactor in a coaxial transmission line. I. Analytical study. *Physics of Plasmas*, 2007, 14:033508.

36. Gusarova MA, Isaev IV, Kaminsky VI. Multipactor simulations in axisymmetric and non-axisymmetric radio frequency structure, Proceedings of RuPAC, 215–217, Zvenigorod, Russia, 2008.

37. Gusarova MA, Kaminsky VI, Kravchuk LV, Kutsaev SV, Lalayan MV, Sobenin NP, Tarasovb SG. Multipacting simulation in accelerating RF structures. *Nuclear Instruments and Methods in Physics Research Section A: Accelerators, Spectrometers, Detectors and Associated Equipment*, 2009, 599(1):100–105.

38. Hockney RW, Eastwood JW. *Computer Simulation Using Particles.* New York: McGraw-Hill, 1981.
39. Eastwood JW. The virtual particle electromagnetic particle-mesh method. *Computer Physics Communications,* 1991, 64(2):252–266.
40. Goplen B, Ludeking L, Smithe D, Warren G. User configurable MAGIC for electromagnetic PIC calculations. *Computer Physics Communications,* 1995, 87(1–2):54–86.
41. Lemke RW, Genoni TC, Spencer TA. Three-dimensional particle-in-cell simulation study of a relativistic magnetron. *Physics of Plasmas,* 1999, 6(2):603.
42. Rasch J, Semenov VE, Rakova E, Anderson D, Johansson JF, Lisak M, Puech J. Simulations of multipactor breakdown between two cylinders. *IEEE Transactions on Plasma Science,* 2011, 39(9):1786–1794.
43. Sazontov A, Anderson D, Vdocicheva N. Simulations of multipactor zones taking into account realistic properties of secondary emission, The 4th International Workshop on Multipactor, Corona and Passive Intermodulation in Space RF Hardware, ESA-ESTEC, Noordwijk, The Netherlands, 2003.
44. Aviviere T, Antonio P, Ming Y. Multipactor breakdown simulation code, The 7th International Workshop on Multipactor, Corona and Passive Intermodulation in Space RF Hardware, Valencia, Spain, 2011.
45. Li Y, Cui WZ, He YN, Wang XB, Hu TC, Wang D. Enhanced dynamics simulation and threshold analysis of multipaction in the ferrite microwave component. *Physics of Plasmas,* 2017, 24(2):023505.
46. Li Y, Wang D, Yu M., He YN, Cui WZ. Experimental Verification of Multipactor Discharge Dynamics Between Ferrite Dielectric and Metal. *IEEE Transactions on Electron Devices,* 2018, 65(10):4592–4599.
47. You JW, Wang HG, Zhang JF, Li Y, Cui WZ, Cui TJ. Highly efficient and adaptive numerical scheme to analyze multipactor in waveguide devices. *IEEE Transactions on Electron Devices,* 2015, 62(4):1327–1333.
48. Wang XB, Li YD, Cui WZ, Li Y, Zhang HT, Zhang XN, Liu CL. Global threshold analysis of multicarrier multipactor based on the critical density of electrons. *Acta Physica Sinica,* 2016, 65(4):047901.
49. Wang HG, Zhai YG, Li JX, Li Y, Wang R, Wang XB, Cui WZ, Li YD. Fast particle-in-cell simulation method of calculating the multipactor thresholds of microwave devices based on their frequency-domain EM field solutions. *Acta Physica Sinica,* 2016, 65:237901.
50. Cui WZ, Zhang H, Li Y, He Y, Wang Q, Zhang HT, Wang HG, Yang J, An improved secondary electrons energy spectrum model and its application in multipactor discharge, *Chinese Physics B,* 2018, 27(3): 038401.
51. Hu TC, Cui WZ, Bao Y, Ma JZ, Cao M. Fabrication and secondary electronic emission property of silver micro-structure by PS sphere. *Metal Materials & Engineering,* 2017, 46(9): 2702–2707.
52. Wu D, Ma J, Bao Y, et al. Fabrication of Porous Ag/TiO$_2$/Au coatings with excellent multipactor suppression. *Scientific Reports,* 2017, 7:43749.

CHAPTER 2

Basic Theory and Measurement Method of Secondary Electron Emission in Multipactor

2.1 OVERVIEW

Secondary electron (SE) emission refers to the phenomenon that electron emission from the surface of a solid material (metal, semiconductor and insulator) when incident electrons (primary electron, PE) or other particles with a certain energy impact on the surface of these objects. As early as 1899, Campbell discovered the phenomenon of secondary electron emission (SEE). In 1902, German scientists Austin and Starke reported the phenomenon of SEE [1]. Since then, the characteristics, mechanism and application of SEE of various substances have been studied in depth.

The secondary electron yield (SEY) and the secondary electron spectrum (SES) are usually used to measure the ability of the SEE. SEY is defined as the ratio of the number of SEs emitted from the surface of the material, including true secondary electrons (TSEs) and backscattered electrons (BEs), to the number of PEs [2]. Figure 2.1 shows a typical curve of SEY σ with incident energy E_{PE}. It can be seen that with increasing E_{PE}, SEY σ increases first and then decreases. In the process of SEY increasing,

DOI: 10.1201/9781003189794-2

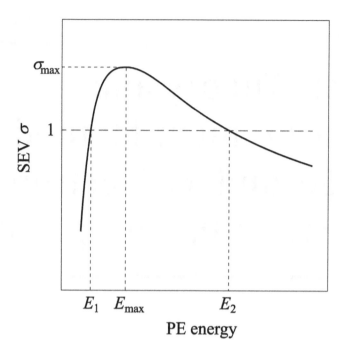

FIGURE 2.1 Typical SEY curve.

when SEY reaches 1 for the first time, the corresponding incident energy E_{PE} is the first cross energy point E_1; when SEY increases to the σ_{max}, the corresponding PE energy is E_{max}; during the decrease, when SEY reaches 1 again, the corresponding PE energy is the second cross energy point E_2. In the fields of accelerator, microwave device and multipacting, the performance of the device and system is very sensitive to the peak value of SEY. Only when the peak value of SEY is small enough, can the device not be damaged by the electron multiplication effect. Therefore, the peak value of SEY is an important parameter in SEY research. In addition, E_1 and E_2 are of great significance to suppress multipactor and reduce discharge probability of materials.

SES is another important factor to describe the energy distribution of SEs. The typical SES curve is shown in Figure 2.2a. As the SEs include the TSEs with lower energy mainly and the high-energy BEs, SEE has an obvious TSE peak at the lower SE energy and a rapidly rising BE peak around the incident energy. Since TSEs account for the majority of the SEs, reflecting the energy distribution of most SEs, the TSE peak of SES is often more concerned. Figure 2.2b shows the TSE peak of SES normalized according

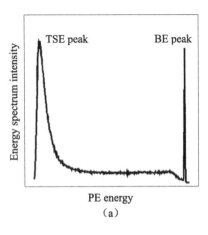
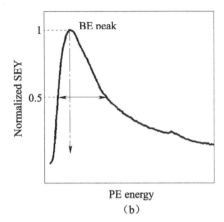

FIGURE 2.2   Typical SES curve. (a) SES, (b) SES TSE peak.

to the maximum value. The study of TSE peaks focuses on two parameters: full width at half maximum (FWHM) and most probable energy (MPE). FWHM is the peak width when the peak value drops to half, reflecting the energy distribution range and concentration area of TSE. MPE is the SE energy corresponding to the peak, reflecting the energy of the largest number of TSE.

2.2 BASIC THEORY OF SEE

2.2.1 Principle of SEE

SEs mainly include TSEs and BEs. The mechanism of SE generation can be summarized as follows (Figure 2.3): when electrons impact upon the surface of a material with a certain energy, it will scatter with atoms or molecules in the material for many times, and some PEs will be elastically scattered by the surface atoms and bounced back directly to form elastic BE. Others entering the material may be in elastically scattered by the material atoms to excite internal SEs. The internal SEs are mainly exited from conduction band, valence band or inner shell electrons due to PEs ionizing and escape from the sample. A part of internal SEs will move to the surface and overcome the work function to form TSE. Due to multiple scattering, some internal SEs change the trajectory and lose energy until they escape from the surface to form inelastic BEs, or stay inside the sample after consuming all the energy, absorbed by the material. Generally, elastic and inelastic BEs are collectively referred to as BE.

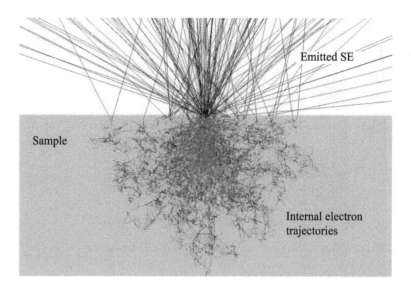

FIGURE 2.3   Schematic diagram of the interaction between PEs and materials.

SE is mainly composed of TSE, while the proportion of BEs is less. In theory, these two kinds of electrons can be classified according to their scattering types, but in experiments, the SEs with energy less than 50 eV are regarded as TSEs, and SEs with energy more than 50 eV are regarded as BEs.

At present, the phenomenon of SEE is an important research topic in the field of surface microanalysis, such as scanning electron microscopy (SEM) [3–5] and Auger electron spectroscopy (AES) [6–8], it is also an important factor restricting the performance of electron accelerator [9–12], high-power microwave source [13–15] and space microwave device [16,17]. In recent years, in aerospace fields, it is necessary to reduce the threshold value of multipactor as much as possible to improve the stability of the device [18–21].

2.2.1.1 Electron Internal Collision

It is inevitable that electrons collide with atoms or molecules in the process of moving inside materials, and this collision will change their trajectories according to some laws, this process is called scattering. The scattering process can be divided into elastic scattering and inelastic scattering according to the criteria whether the energy is lost or not. Elastic scattering is the scattering of electrons by Coulomb potential of nucleus and electron cloud. As the mass of the nucleus is much larger than that of the electron (more than three orders), the energy changes of the atom and the PE can be ignored after scattering, and only the motion direction of the PE has changed. In the process of inelastic scattering, not only the motion direction of PEs is changed but also the energy is lost.

Figure 2.4 shows the difference between elastic and inelastic scattering. For the elastic scattering of electrons by atoms without energy exchange, only the movement direction of the electrons is changed. While for inelastic scattering, the energy lost by PEs is transferred to the extra nuclear electrons, which are released from the nucleus, and an internal SE is excited to form a vacancy.

In order to analyse the scattering process of electrons, it is necessary to calculate the scattering cross-section of electrons. The scattering cross-sections of electrons can be divided into elastic and inelastic scattering cross-sections according to different types of scattering. The elastic scattering cross-section of the PE is the integral of the differential elastic scattering cross-section of the PE in all directions, while the inelastic scattering cross-section of the PE is the integral of the differential inelastic scattering cross-section of the PE in all directions and various energy losses. Among them, the differential elastic scattering cross-section is the probability $d\sigma_e/d\Omega$ that the PEs are elastically scattered into a unit solid angle in a certain direction. The total elastic scattering cross-section is the integral of the differential elastic scattering cross-section of the PEs in all direction, it can be expressed as

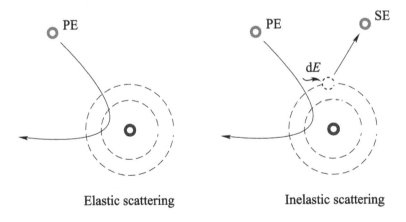

Elastic scattering   Inelastic scattering

FIGURE 2.4   Schematic diagram of the electron scattering by material atoms.

$$\sigma_e = \int \frac{d\sigma_e}{d\Omega} d\Omega \qquad (2.1)$$

and the inelastic scattering cross-section of an atom σ_{in} is the double integral of differential inelastic scattering $\sigma'_{in}(\Omega, E_f)$ of an atom in all solid angles and all PEs in the final state energy range E_f, that is

$$\sigma_{in} = \iint \sigma'_{in}(\Omega, E_f) d\Omega \, dE_f \qquad (2.2)$$

2.2.1.2 Electron Emission

When electron enters the sample, it will collide (scatter) with the sample atoms many times, changing the direction of motion and losing energy to escape from the sample surface or stay in the sample until the energy is exhausted. The process is shown in Figure 2.5, where θ is the scattering

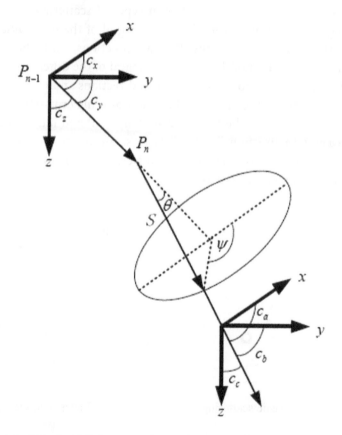

FIGURE 2.5   Relationship between electron scattering trajectory and coordinate system.

angle, ψ is the azimuth angle, (c_x, c_y, c_z) and (c_a, c_b, c_c) represent the cosine value of the angle between direction of electron motion and x, y, and z axis before and after scattering, respectively.

After each collision, the electron can reach any point on the circle where the azimuth angle is located, so the azimuth angle is evenly distributed in the range of 0~2π, which can be expressed as:

$$\psi = 2\pi \times R \qquad (2.3)$$

where R is a random number uniformly distributed on the interval (0, 1). With the scattering angle and azimuth angle, the cosine value of the angle along the coordinate axis after collision can be calculated as

$$\begin{cases} c_a = c_x \times \cos\theta + V_1 \times V_3 + c_y \times V_2 \times V_4 \\ c_b = c_y \times \cos\theta + V_4 \times (c_z \times V_1 - c_x \times V_2) \\ c_c = c_z \times \cos\theta + V_2 \times V_3 - c_y \times V_1 \times V_4 \end{cases} \qquad (2.4)$$

where V_1, V_2, V_3, and V_4 are temporary variables, and satisfy the following relationship

$$\begin{cases} V_1 = A_N \times \sin\theta \\ V_2 = A_N \times A_M \times \sin\theta \\ V_3 = \cos\phi \\ V_4 = \sin\phi \end{cases}, \qquad (2.5)$$

The expressions of A_M and A_N are as follows:

$$\begin{cases} A_M = -(c_x/c_z) \\ A_N = 1/\sqrt{1 + A_M^2} \end{cases} \qquad (2.6)$$

After the relationship between the direction of electron movement and the coordinate axis is determined, it is easy to get the next position of the electron

$$\begin{cases} x_n = x + S \times c_a \\ y_n = y + S \times c_b \\ z_n = z + S \times c_c \end{cases} \qquad (2.7)$$

among them,

x, y, z – initial coordinates before electron scattering;

x_n, y_n, z_n – the position coordinates of the electron arrival after scattering.

2.2.1.3 Influence of Surface Barrier

Electron emission from the material needs to cross the barrier region of the interface. In order to calculate the movement of electrons in the barrier region, we need to consider their wave property through the method of quantum mechanics. Assuming that, by Monte Carlo simulation, when an electron reaches the boundary, its energy is E_{in}, the angle between its direction and the normal of the interface is α, the distribution of the potential barrier of the interface along the normal of the interface is $U(x)$, then the wave function of the electron ψ satisfies the Schrodinger equation:

$$\frac{\hbar^2}{2m}\frac{\partial^2\psi(x)}{\partial x^2}+\left(E_{in}\cos^2\alpha-U(x)\right)\psi(x)=0, \qquad (2.8)$$

where $E_{in}\cos^2\alpha$ represents the vertical component of the SE energy. Only the vertical component of the SE energy should be considered because the movement of electrons in parallel directions does not affect their transmission.

In the material, the barrier does not affect the energy and motion of electrons. Assuming that the potential energy is constant, the solution is a uniform plane wave

$$\psi_m(x)=A_m e^{ik_m x}+B_m e^{-ik_m x}$$
$$\psi_v(x)=A_v e^{ik_v x}+B_v e^{-ik_v x}, \qquad (2.9)$$

where subscript m and v represent to the area within the material and the microtrap, respectively. The wave vector of electrons in the two regions is $k_m=\sqrt{2mE_{in}\cos^2\alpha}\big/\hbar$ and $k_v=\sqrt{2m\left(E_{in}\cos^2\alpha-U_v\right)}\big/\hbar$, where m is the electronic mass, and $\hbar$ is the reduced Planck constant. The coefficients A and B are the amplitude of the incident and reflected waves, respectively.

The distribution of barriers $U(x)$ is a function of position x in the barrier region, so it is difficult to obtain the analytical solution of the wave function. For this reason, the potential barrier is divided into n layers, and the barrier is considered to be a constant U_i, $1 \leq i \leq n$ if the thickness of each layer is small enough. Therefore, the wave function in each layer has a general solution as follows,

$$\psi_i(x) = A_i e^{ik_i x} + B_i e^{-ik_i x}$$

$$k_i = \sqrt{2m\left(E_{\text{in}} \cos^2 \alpha - U_i\right)} / \hbar,$$

(2.10)

ψ_i and k_i represent the wave function and wave vector in the ith layer, respectively. In addition, the metal may be referred to as the 0th layer, and the vacuum outside the barrier may be referred to as the $n + 1$th layer. The wave function satisfies the continuity boundary condition at the interface between layers

$$\psi_i(x_i) = \psi_{i+1}(x_i)$$

$$\left. \frac{\partial \psi_i}{\partial x} \right|_{x=x_i} = \left. \frac{\partial \psi_{i+1}}{\partial x} \right|_{x=x_i},$$

(2.11)

where x_i is the position of the interface between the ith layer and the $i + 1$th layer. The SE will not return to the metal after entering the vacuum, that is, there is no reflected wave in the vacuum, $B_{n+1} = 0$. The relationship between A_0 and A_{n+1} can be obtained by solving the above formula. The square of the amplitude of the wave function represents the density of electronic states, so the transmission coefficient T is

$$T = \frac{k_{n+1} |A_{n+1}|^2}{k_0 |A_0|^2},$$

(2.12)

There is a T probability for electron to cross the interface barrier, and the electrons passing the barrier satisfy the law of conservation of energy and momentum

$$E(0) + U(0) = E(h) + U(h)$$

$$p_\parallel(0) = p_\parallel(h)$$

(2.13)

$$\frac{p_\perp^2(0)}{2m} + U(0) = \frac{p_\perp^2(h)}{2m} + U(h),$$

where $p_{\parallel}$ and $p_{\perp}$ represent the electron momentum parallel and perpendicular to the exit surface, respectively. The movement of electrons in the microsize trap structures can be analysed by particle tracking. For microstructure in metal that is an equipotential body, the internal space of the structure is an area that is almost surrounded by equipotential surfaces and also very close to the equipotential body. The movement of electrons in the microstructure can be regarded as linear movement. However, for the microstructure in the medium, there is a certain electric field due to the influence of the medium charging, the movement of electrons is affected by the electric field. The electric field in the microstructure can be obtained by the finite difference or finite element method. The trajectory of electrons satisfies the following equation

$$m\ddot{x}(t) = -eE_X(x,z,t),$$
$$m\ddot{z}(t) = -eE_Z(x,z,t). \tag{2.14}$$

which can be solved by Runge-Kutta methods.

The energy and angle of electron emission are obtained by analysing the interactions between electrons and the surface. The energy spectrum distribution of SEs is obtained by counting the energy distribution of a large number of electrons.

Next, we need to determine whether these SEs can cross the surface barrier. As shown in Figure 2.6, for an ideal surface, the internal electrons

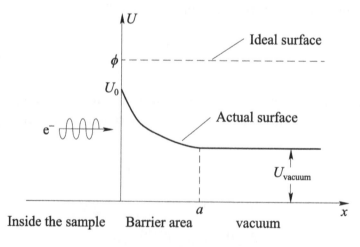

FIGURE 2.6   Schematic diagram of SEs crossing the surface barrier.

only need to overcome the work function of the metal to escape, while for a non-ideal surface, there is a surface barrier region between the metal and the vacuum, where the vacuum level distorted, and the final barrier V may be higher or lower than U. Given the surface barrier distribution, the wave functions of the electron with energy E_{in} and angle θ satisfy the Schrödinger equations within the sample, the barrier and the vacuum, thus the density of electronic states in the three regions can be calculated, and the transmission coefficient $T(E_{in}, \theta)$ can be obtained according to the ratio of the density of electronic states on both sides of the barrier.

2.3 SIMULATION OF SEE

The basic research on SE emission is widely used, including promotion and suppression. Examples of promoting SEE phenomena include: electron multipliers, material surface composition, structure analysis (using transmission electron microscope, scanning electron microscope, Auger electron spectrometer, electron diffractometer, etc.), solid electrostatic protection, plasma display and so on. In other fields, SEE is undesired, i.e., the electron cloud in t particle accelerators, the multipactor of high-power microwave devices in space, the multistage depressed collector efficiency of traveling wave tubes, the surface charging on spacecraft, the breakdown of dielectric window in high-power microwave source and so on.

As mentioned above, because of the wide application of SEE, the research on this phenomenon is still widely concerned. There are two methods to describe SEE process of materials in the published literature: the Monte Carlo simulation method based on multiple scattering processes of electrons in solid and the phenomenological probability method based on experimental data fitting. Monte Carlo is a simulation calculation method based on actual physical processes, calculating elastic and inelastic scattering cross-sections, tracking electron motion trajectories and simulating energy decay processes; the phenomenological probability method describes the SEE characteristics based on measured data by fitting appropriate formulas and parameters, and is characterized by the simplicity of physical processing and the accuracy of experimental data fitting.

2.3.1 Theoretical Formula

2.3.1.1 Furman Model

The phenomenological probability model of SEE mainly solves three problems: the number of exited SEs, the energy and direction of emitted SEs.

When the initial electrons with a certain energy impact on the material, three kinds of SEs may be exited: elastic BE, inelastic BE and TSE. For a certain collision, which kind of SE is excited is a random event with a certain probability. The probability of elastic backscattering is:

$$P_e\left(E_p,\theta_p\right)=\varepsilon\left(E_p,\theta_p\right) \tag{2.15}$$

where

$$\varepsilon\left(E_p,\theta_p\right)=\varepsilon\left(E_p\right)^{\cos\theta_p}\cdot C_2^{1-\cos\theta_p}$$

$$C_2=\chi\cdot\frac{\varepsilon\left(E_p\right)}{\varepsilon\left(E_p\right)+\eta\left(E_p\right)}$$

$$\varepsilon\left(E_p\right)=\frac{\varepsilon_1}{1+E_p/E_{e1}}+\frac{\varepsilon_2}{1+E_p/E_{e2}}$$

where $\varepsilon_1 = \varepsilon_0 - \varepsilon_2$, ε_0 is the elastic backscattering coefficient when the initial electron incident energy is zero and usually takes the value 1; E_p is the incident energy of PE; θ_p is the incident angle of PE; $\varepsilon_2 = 0.07$, $E_{e1} = g/\sqrt{Z}$, $E_{e2} = h\cdot Z^2$, $g = 50$, $h = 0.25$, Z is the atomic number, and $\chi = 0.89$.
The probability of inelastic backscattering is:

$$P_i\left(E_p,\theta_p\right)=\eta\left(E_p,\theta_p\right) \tag{2.16}$$

where

$$\eta\left(E_p,\theta_p\right)=\eta\left(E_p\right)^{\cos\theta_p}\cdot C_1^{1-\cos\theta_p}$$

$$C_1=\chi\cdot\frac{\eta\left(E_p\right)}{\varepsilon\left(E_p\right)+\eta\left(E_p\right)}$$

$$\eta\left(E_p\right)=a\cdot\left(1-b\cdot E_p\right)\cdot E_p^\gamma\cdot\exp\left(-\left(\frac{E_p}{E_b}\right)^\mu\right)$$

where $E_b = c + d\cdot Z$, a depends on the material properties, and the value usually is $\left[7\times10^{-3}, 10\times10^{-3}\right]$ (the value in this chapter is 0.0078), $b = 3.0\times10^{-5}$, $c = 300$, $d = 175$, $\gamma = 0.56$ and $\mu = 0.70$.

Only one type of electron emission can occur in one collision, such as elastic backscattering, inelastic backscattering and TSE emission. Therefore, the probability of the TSE emission is as follows:

$$P_s\left(E_p,\theta_p\right)=1-\varepsilon\left(E_p,\theta_p\right)-\eta\left(E_p,\theta_p\right) \tag{2.17}$$

The probability of three kinds of SEE is determined, the number, energy and direction of emitted electrons can be calculated. For elastic backscattering or inelastic backscattering, the SEY is 1. For the case of TSE emission, the number of emitted electrons can be infinity theoretically. According to experimental results, Poisson distribution function can be used to describe the number of the emitted SEs:

$$P_n\left(E_p,\theta_p\right)=\frac{e^{-\lambda}\cdot\lambda^n}{n!}, \tag{2.18}$$

where n is the number of TSEs and P_n is the probability of the emitted number n. In the programming of calculation, the upper limit of n is determined according to the principle that the sum of probabilities is 1 (taken as 120 in this chapter). The expected value λ is:

$$\lambda\left(E_p,\theta_p\right)=\frac{\delta\left(E_p,\theta_p\right)}{1-\varepsilon\left(E_p,\theta_p\right)-\eta\left(E_p,\theta_p\right)}, \tag{2.19}$$

TSE emission coefficient at normal incidence is:

$$\delta\left(E_p\right)=\delta_m\cdot\frac{s\cdot E_p/E_m}{s-1+\left(E_p/E_m\right)^s} \tag{2.20}$$

where δ_m, E_m and s are constant related to the material. For oblique incidence, the energy of incident electron corresponding to the maximum SEY E_m is:

$$E_m\left(\theta_p\right)=E_{m0}/\cos^n\theta_p, \tag{2.21}$$

where r_1 is a constant related to material and the surface.
For oblique incidence, TSE emission coefficient is:

$$\delta\left(E_p,\theta_p\right)=\delta\left(E_p\right)\cdot\frac{k+1}{k+\cos\theta_p} \tag{2.22}$$

where $k = p \cdot Z + r$, $p = 0.0027$, r is a constant related to surface: $r = 0$, for clean and smooth surfaces, for rough surfaces $2.5 < r < 10$, typically $r = 5$.

Next, we calculate the energy of SEs: for elastic backscattering, the energy of SEs is equal to the incident energy of PEs. For inelastic backscattering, the energy of SEs is:

$$E_b = E_p \cdot G_b(u) \tag{2.23}$$

where the probability density function is:

$$G_b(u) = \alpha^{-1/n_b} \left(\arccos \left(1 - \beta \cdot u \right) \right)^{1/n_b}, \tag{2.24}$$

where n_b, X_{cb} are undetermined parameters and are usually taken as $n_b = 1.5$, $X_{cb} = 0.9$, $\beta = 1 - \cos\alpha$, $\alpha = \pi \cdot X_{cb}^{n_b}$ and u is a random number uniformly distributed in the interval $(0,1)$. The emission energy of TSE is:

$$E_{se} = E_{re} \cdot G_s(u) \tag{2.25}$$

where for the first TSE, $E_{re} = E_{re,1} = E_p$; for the following TSEs, $E_{re,i} = E_{re,i-1} - E_{se,i-1}$; $E_{re,i}$ represents the energy when the ith TSE is emitted; $E_{se,i-1}$ represents the energy of the $(i-1)$th TSE. In this way, the total energy of the emitted electrons will not exceed the energy of the incident electrons, thus ensuring the law of conservation of energy.

The energy probability density function of TSE is:

$$G_s(u) = \left(\frac{2}{\pi} \cdot \tan^{-1} \left(\sqrt{\tan\left(\frac{\pi}{2} \cdot X_{cs} \right) \cdot \tan\left(\frac{\pi}{2} \cdot u \right)} \right) \right)^{1/n_s}, \tag{2.26}$$

$$X_{cs} = X_c / \left(0.9 + 1.1 \cdot X_c \right)$$

$$X_c = 4 \cdot \left(B - e^{-E_{tr}/A} \right) / E_{tr}$$

$$E_{tr} = \begin{cases} E_{re}, & E_{re} > 1\,\text{eV} \\ 1, & E_{re} \leq 1\,\text{eV} \end{cases}$$

where n_s, A and B are undetermined parameters related to the material and usually taken their values as $n_s = 0.51$, $A = 4.0$ and $B = 5.8$.

As for the simulation of emitted direction of SEs, assuming that the emitted directions of elastic and inelastic backscattering are determined according to the specular reflection law; the emitted polar angle of TSEs follows the cosine distribution and the azimuth angle follows the uniform distribution. It should be noted that, in the actual emission process, the inelastic BEs are excited by PEs entering the metal material, so their emitted direction does not meet the specular reflection law in theory, this problem will be further explained in Chapter 3, the simulation results of the SEY in the microtrap structure. The assumptions above are only for the convenience of calculation. Figure 2.7 shows the fitting results of SEY varying with incident angle and energy on the silver-plated aluminium alloy surface using the above phenomenological probability method. The calculated results are in good agreement with the measured results, which proves the feasibility of using the phenomenological probability method to simulate the SEY of the microtrap structure. The parameters of the phenomenological probability model obtained are as follows: $\delta_m = 1.95$, $E_{m0} = 280$, $r_1 = 0.5$, $s = 1.7$, $r = 2$, $\chi = 0.89$, and other unspecified parameters are the same as mentioned above. It should be noted that the SEY of metal surface usually changes with the time in the atmosphere because of

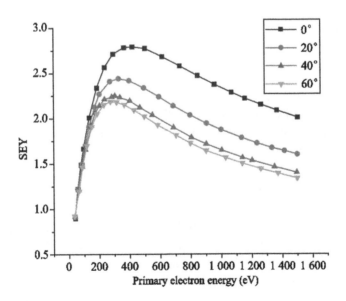

FIGURE 2.7 Calculation result of phenomenological probability of smooth silver-plated surface.

the influence of surface contamination and adsorption. The experiment results in the figure are measured on silver-plated surface which has been stored for a long time (more than half a year) in the atmosphere. The SEY curve described by the probability model of the silver-plated surface is slightly smaller than that explained here, which is due to the short storage time of the sample in the atmosphere, and the SEY has not yet reached a stable state.

2.3.1.2 Everhart Model

The Chung-Everhart (C-E) model is based on the fact that PEs perpendicularly incident on the semi-infinite metal surface. The PEs collide with the metal material to produce SEs at a certain depth. Ignoring the collision inside the metal, SE with energy E' are emitted from the surface obliquely. The C-E model describes the energy distribution of TSEs. The Fermi energy E_F and work function Φ on the emitted electrons are considered. The basic condition for electrons to be emitted from the surface is $E' \geq E_F + \Phi$. Based on the physical process of electron collision, the C-E model is an expression derived from the physical theory involved in the process from electrons incidence to emit, so the parameters of the CE model have its physical meaning.

$$\frac{dN}{dE'} = \frac{e^4 k_F^3 14.7 \left(E_F \beta\right)^{3/2} \left(E' - E_F - \Phi\right)}{12\pi E_p \left(m^*\right)^{1/2} \left[\tan^{-1}(1/\beta)^{1/2} + \beta^{1/2}/(1+\beta)\right]\left(E' - E_F\right)^4}, \quad (2.27)$$

where $\beta = (4/9\pi)^{1/3} \left(\gamma_s/\pi\right)$, γ_s represents the radius of electron, m^* is the effective mass of electron, E_0 is the PE energy and k_F is the Fermi wave vector. The shape of dN/dE', which can be seen from the expression, is completely determined by the following factors

$$f(E') = (E' - E_F - \Phi)/(E' - E_F)^4, \quad (2.28)$$

The multiplier affects the magnitude of dN/dE' only. The peak value and FWHM are determined by $f(E')$. The maximum of dN/dE' occurs when the SE energy is E'_m, i.e., $f(E')$ reaches the maximum. It is easy to get this value by differentiating the expression

$$E'_m = E_F + \frac{4}{3}\Phi, \quad (2.29)$$

The FWHM of dN/dE' can also be obtained from $f(E')$. For simplification, assuming the energy of SE emitted is E, then

$$E = E' - E_F - \Phi, \tag{2.30}$$

The expression $f(E)$ describes the trend of the model, while the parameter k represents the other parameters,

$$\frac{dN}{dE} = \frac{kE}{(E+\Phi)^4}, \tag{2.31}$$

This is a simplified C-E model, except for the normalized parameter k, the only variable is the work function Φ, so the work function has an important influence on the energy spectrum. Obtaining the actual work function on the material surface is of great significance for accurate simulation of energy spectrum. The specific material has a certain work function, however, the cleanliness of the material surface, the thickness of the surface adsorption layer, and the surface structure will vary with the work function, so fitting the energy spectrum is mainly adjusted by the parameters related to the work function.

It can be seen from the fitting energy spectrum in Figure 2.8 that the fitting results of the C-E model in the range of 0–50 eV are acceptable, mainly by adjusting parameters related to the work function, and the normalized parameter k has little influence. According to the authoritative data of material work function, the work function of Cu, Ag and Au is 4.65, 4.26 and 5.1, respectively. Compared with the fitting data, the work functions of Cu and Au are close to the fitted data, while the data of Ag are different. In addition, the normalized fitting parameters of Au in different energy range are different because the PE energy E_0 is also included in the normalized parameters in the simplified model.

2.3.1.3 Semi-Physical Model

SEs are the electrons emitted from the surface of materials, including TSEs and backscattered secondary electrons (BSEs). The TSEs are produced by the inelastic scattering of the PEs and the extranuclear electron, while the BSEs are directly emitted from the elastic collision between the PEs and the nucleus. There are many ways to calculate the SEE, including Monte Carlo simulation and fitting calculation. Monte

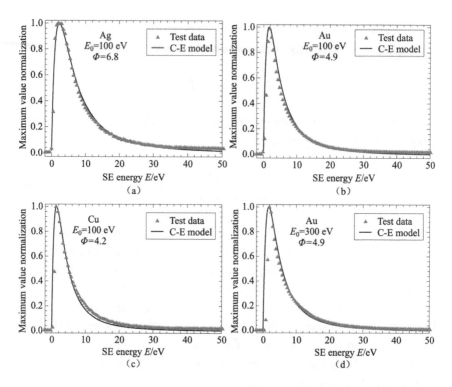

FIGURE 2.8   (a–d) Fitting results of C-E model and experimental test data.

Carlo simulation tracks every collision and movement of electrons, which is slow and inaccurate. These problems can be avoided by fitting experimental data, such as Joy, Furman and Vaughan formula. Among them, Joy and Vaughan formulas are inconsistent with the experimental data in higher PE energies, while the Furman model is a fitting formula which completely depends on the experimental data and with great limitations. Therefore, it is necessary to build a more appropriate model for complex surface conditions. A comprehensive semi-physical model is proposed to calculate the SEE of metals, calculating the excitation, movement and emission of SE, and the oblique incidence and surface contamination are considered.

2.3.1.3.1 SE Excitation   When an electron with a certain energy enter the metal material and are scattered, internal SEs will be excited within the range of its incidence depth. Therefore, the electron range of the PEs in material is a parameter closely related to the SEE.

TABLE 2.1   The Density of Commonly Used Aerospace Metal Materials

Material Name	Atomic Number (Z)	Density (g cm⁻³)
Au	79	19.30
Ag	47	10.49
Cu	29	8.96
Al	13	2.70

For amorphous materials, the relationship of the electron range $R(E_{pe})$ and PE energy is

$$R(E_{pe}) = \frac{L_k}{\rho_r}\left(\frac{E_{pe}}{E_k}\right)^\alpha,$$ (2.32)

where E_{pe} is the energy of the PE, the value of E_k is 1 keV, ρ_r is the density of the material relative to water, L_k is the Lane-Zaffarano constant and its value is 76 nm. The density of metal materials commonly used in aerospace is shown in Table 2.1.

Equation (2.32) can also be expressed as

$$\ln\left(\frac{R(E_{pe})}{L_k}\right) = \alpha\ln\left(\frac{E_{pe}}{E_k}\right) - \ln\rho_r.$$ (2.33)

It can be seen that α reflects the penetration ability of electrons in the material. For the PE with a given energy, the larger the value of α, the larger the electron range $R(E_{pe})$ and the stronger for electrons to penetrate the material. Conversely, the smaller the α, the weaker for electrons to penetrate the material. Lane and Zaffarano believed that α is a constant independent of the material. This conclusion is in good agreement at higher PE energies (> several keV), but it is not accurate at lower PE energies. In this chapter, we analyse the data from NIST, simulate the electron scattering process and find that,

1. The larger the atomic number of the material, the larger angular deflection of electrons elastic scattered in the material, resulting in the weaker penetration ability of electrons.

2. Materials with high atomic number have more extranuclear electrons. The energy loss of PE will reduce the penetration of the electrons, due to the energy exchange with extranuclear electrons.

In this chapter, it is found that the coefficient α in the electron range is related to the material, through quantitative analysis of aluminium, gold, silver, copper and so on.

$$\alpha = 1.26 + 0.46\exp(-Z/19.92), \tag{2.34}$$

As shown in Figure 2.9, PEs undergo a series of inelastic scattering in the material and will excite internal PEs within the electron range. Strictly, the number of internal SEs excited varies with the depth of the PE range, that is to say, the density of internal SEs will have a certain distribution in the depth. However, the internal SEs are nearly uniformly distributed in the range of escape depth which is closely related to the SEE, so the following approximation can be made

$$n(z) = \frac{N\left(E_{\mathrm{pe}}\right)}{R\left(E_{\mathrm{pe}}\right)}, \tag{2.35}$$

where $R\left(E_{\mathrm{pe}}\right)$ is the range of PEs and $N\left(E_{\mathrm{pe}}\right)$ is the total number of internal SEs.

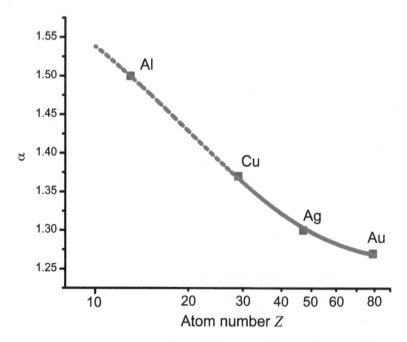

FIGURE 2.9   Relationship between coefficient α and atomic number.

2.3.1.3.2 Internal SEE   Internal SEs excited by PEs will collide with free electrons and lattice atoms to lose energy in the material. After several collisions, if the internal SEs cannot reach the surface of the material, they may be trapped by the material atoms to be kept in the material because of the loss energy. Therefore, only a part of the SEs can move to the surface of the material. Bruining [22] and Wittry et al. [23] simplified the direction of movement of internal SEs into upward and downward and believed that the probability of internal SEs at depth z moving to the surface of the material is

$$p(z) = \frac{1}{2}\exp\left(-\frac{z}{\lambda_{\mathrm{eff}}}\right), \qquad (2.36)$$

where λ_{eff} is the effective escape depth of internal SEs.

The actual movement direction of the internal SEs is not simply downward or upward but in all directions. For electrons moving in a specific direction, the probability of being absorbed by the distance travelled ds is $p_{\mathrm{abs}}(ds)$, which is proportional to the distance ds:

$$p_{\mathrm{abs}}(ds) = \gamma ds, \qquad (2.37)$$

where γ is a proportional constant, representing the absorption coefficient. According to the above formula, the probability that the internal SEs are not absorbed in the distance ds is:

$$f(ds) = 1 - p_{\mathrm{abs}}(ds) = 1 - \gamma ds, \qquad (2.38)$$

Assuming the probability that the electron is not absorbed through the distance s is $f(s)$, and the probability that the electron is not absorbed through the distance $s + ds$ is $f(s + ds)$, then according to the probability theory,

$$f(s + ds) = f(s)f(ds) = f(s)(1 - \gamma ds), \qquad (2.39)$$

or

$$\frac{f(s + ds) - f(s)}{ds} = -\gamma f(s), \qquad (2.40)$$

that is, the following differential equation about $f(s)$

$$\frac{df(s)}{ds} = -\gamma f(s).$$ (2.41)

Obviously, the physical meaning of $f(0)$ is the probability that the electron is not absorbed at the initial moment, and its value is 1. Combining differential equations and initial conditions, it is obtained as

$$f(s) = \exp(-\gamma s),$$ (2.42)

In this chapter, it is considered that the motion direction of the internal SE is shown in Figure 2.10. For the internal SE at depth z and with an angle φ between the motion direction and the outer surface of the material, the distance it travels to the surface is $z / \cos\varphi$.

The emission probability of internal SEs decays exponentially, and the probability of internal SEs moving to the material surface in the direction φ is

$$p(z,\varphi) = \exp\left(-\frac{z}{\lambda\cos\varphi}\right),$$ (2.43)

where λ is the average free path of elections move in the material. The λ values of metal materials commonly used in aerospace are shown in Table 2.2.

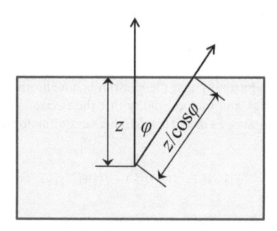

FIGURE 2.10 The distance of internal SEs motion to the surface from different directions.

TABLE 2.2   The Average Free Path of Commonly Used Metal Materials

Material Name	Atomic Number (Z)	Average Free Path of the Election Motion (λ nm^{-1})
Au	79	1.6
Ag	47	1.4
Cu	29	1.2
Al	13	3.8

When the internal SEs move in the material, they may also generate new internal SEs through non-scattering with the atoms, which is called cascade scattering. In the isotropic amorphous materials, the motion direction of the SEs after cascade scattering can be considered isotropic. As shown in Figure 2.11, the proportion of SEs with motion direction $\varphi \sim \varphi + d\varphi$ is represented as the proportion of grey-shaded ring. Assuming the radius of the sphere is 1, the radius of the ring is $\sin\varphi$ and the width of the ring is $d\varphi$, so that the area of the ring is $2\pi \sin\varphi d\varphi$. The surface area of

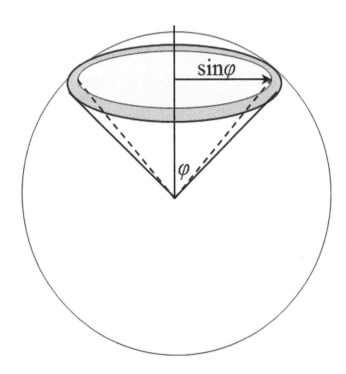

FIGURE 2.11   Schematic diagram of the proportion of internal SEs moving in the direction φ.

the sphere is 4π. The density of internal SEs at the depth z is $n(z)$, and the number of internal SEs in the direction $\varphi \sim \varphi + d\varphi$ is

$$n(z,\varphi)d\varphi = \frac{1}{2}n(z)\sin\varphi d\varphi, \tag{2.44}$$

The number of internal SEs moving to the surface in the direction of $\varphi \sim \varphi + d\varphi$ at the depth of $z \sim z + dz$ is

$$n(z,\varphi)p(z,\varphi)dzd\varphi = \frac{1}{2}n(z)\sin\varphi \exp\left(-\frac{z}{\lambda\cos\varphi}\right)dzd\varphi \tag{2.45}$$

The number of internal SEs moving to the surface in all directions at the depth of $z \sim z + dz$ is

$$n_s(z)dz = \frac{1}{2}n(z)\int_0^{\frac{\pi}{2}}\sin\varphi \exp\left(-\frac{z}{\lambda\cos\varphi}\right)d\varphi dz, \tag{2.46}$$

Integrate the angle first

$$\int_0^{\frac{\pi}{2}}\sin\varphi \exp\left(-\frac{z}{\lambda\cos\varphi}\right)d\varphi = -\int_0^{\frac{\pi}{2}}\exp\left(-\frac{z}{\lambda\cos\varphi}\right)d(\cos\varphi)$$

$$= \int_0^1 \exp\left(-\frac{z}{\lambda y}\right)dy$$

$$= \exp\left(-\frac{z}{\lambda}\right) - z\Gamma\left(0,\frac{z}{\lambda}\right), \tag{2.47}$$

where Γ is the Gamma function $\Gamma(a,z)=\int_z^\infty t^{a-1}e^{-t}\,dt$. Obviously, only the internal SEs moving towards the surface can emit through the material surface, so the integral interval of the angle φ is $[0\ \pi/2]$.

The equation of internal SE density is substituted into equation (2.46) to obtain the contribution of the internal SEs at the depth of $z \sim z + dz$:

$$n_s(z)dz = \frac{N(E_{pe})}{2R(E_{pe})}\left(\exp\left(-\frac{z}{\lambda}\right) - z\Gamma\left(0,\frac{z}{\lambda}\right)\right)dz, \tag{2.48}$$

The total number of SEs at the surface is the sum of all internal SEs moving to the surface within the range of the PEs:

$$N_s = \int_0^{R(E_{pe})} n_s(z) dz$$

$$= \frac{N(E_{pe})}{2R(E_{pe})} \int_0^{R(E_{pe})} \left(\exp\left(-\frac{z}{\lambda}\right) - z\Gamma\left(0, \frac{z}{\lambda}\right) \right) dz$$

$$= \frac{N(E_{pe})\lambda}{4R(E_{pe})} F_\Gamma\left(\frac{R(E_{pe})}{\lambda} \right), \tag{2.49}$$

where

$$F_\Gamma(x) = 2 \int_0^x \left(\exp(-x') - z\Gamma(0, x') \right) dx' \tag{2.50}$$

It cannot be expressed by elementary functions but can be approximated and expressed as

$$F_\Gamma(x) \approx 1 - \exp(-1.6x), \tag{2.51}$$

The difference is shown in Figure 2.12.

Thus, the number of internal SEs at the surface is

$$N_s = \frac{N(E_{pe})\lambda}{4R(E_{pe})} \left(1 - \exp\left(-1.6 \frac{R(E_{pe})}{\lambda} \right) \right), \tag{2.52}$$

The SEs can emit from the surface is dependent on the energy of the SEs. In the work of Joy et al., the energy of internal SEs is considered to be the average ionization energy. Streitwolf et al. considered that the energy distribution of the inner SE satisfies

$$S(E_s) \propto \frac{1}{(E_s - E_F)^2} \tag{2.53}$$

This conclusion has a physical essential flaw: when calculating the energy of all electrons greater than E_F, the integral does not converge, that is, the number of internal SEs tends to infinity, which obviously contradicts the

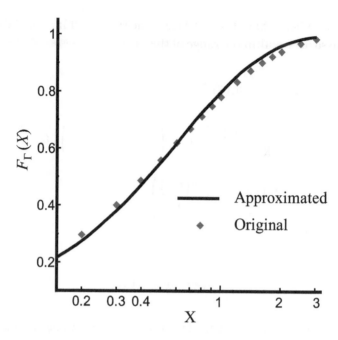

FIGURE 2.12 Comparison of approximate and actual values.

physical reality. This conclusion only considers the process that the PEs excite the internal SEs and ignores the cascade scattering process of the internal SEs.

In this chapter, the cascade scattering process of internal SEs is considered, and the energy of the internal SEs satisfies the exponential distribution:

$$S_R(E_s) = A\exp\left(-\frac{E_s - E_F}{E_v}\right), \qquad (2.54)$$

where E_s is the energy of the internal SE, E_v is the expected value of the energy of the internal SE, and E_F is the Fermi level of the material. $S_R(E_s)dE_s$ represents the probability that the energy of the internal SE is in the range $E_s + dE_s$. The coefficient A satisfies the following normalization conditions

$$\int_{E_F}^{E_{pe}} S_R(E_s)dE_s = 1, \qquad (2.55)$$

then

$$A = \frac{1}{E_v \left(1 - \exp\left(-\dfrac{E_{pe} - E_F}{E_v}\right)\right)}. \qquad (2.56)$$

According to the conservation of energy, the sum of all internal SE energies should be equal to the PE energy. The number of SEs within the energy interval $E_s \sim E_s + dE_s$ is $N(E_{pe})S(E_s)dE_s$. The starting point of the energy of the internal SE is selected at the bottom of the conduction band, that is, the energy obtained by internal SE is the part where E_s greater than the Fermi level E_F of the material, so the energy obtained by the internal SE is $N(E_{pe})S(E_s)(E_s - E_F)dE_s$. The energy obtained by all internal SEs should be equal to the total energy of the PEs, i.e.,

$$N(E_{pe}) \int_{E_F}^{E_{pe}} S_R(E_s)(E_s - E_F)dE_s = E_{pe}, \qquad (2.57)$$

where the maximum energy of SEs is the PE energy E_{pe}, so the upper limit of integration is selected as E_{pe}. Solve the formula (2.57), it can be obtained as

$$N(E_{pe}) = \frac{E_{pe}}{E_v + \dfrac{\exp\left(-\dfrac{E_{pe} - E_F}{E_v}\right)(E_{pe} - E_F)}{1 - \exp\left(-\dfrac{E_{pe} - E_F}{E_v}\right)}}, \qquad (2.58)$$

The internal SEs need to cross the surface barrier to emit from the surface. The emission probability depends on the surface barrier and the kinetic energy of the SEs. The probability of an internal SE with energy E_s crossing the surface barrier U is

$$p_e(E_s) = \begin{cases} 0 & E_s \leq U \\[2mm] \dfrac{4\sqrt{1 - U/E_s}}{\left(1 + \sqrt{1 - U/E_s}\right)^2} & E_s > U \end{cases}, \qquad (2.59)$$

This can be simplified as the following approximate expression

$$
f_0(E_s) = \begin{cases} 0 & E_s \leq U \\ \dfrac{29(E_s/U - 1)}{1 + 29(E_s/U - 1)} & E_s > U \end{cases}, \tag{2.60}
$$

The average emission probability of SEs at the surface is

$$
\begin{aligned}
P_s &= \int_0^{E_{pe}} S_R(E_s) p_e(E_s) dE_s \\
&= 4A \int_U^{E_{pe}} \exp\left(-\frac{E_s - E_F}{E_v}\right) \cdot \frac{\sqrt{1 - U/E_s}}{\left(1 + \sqrt{1 - U/E_s}\right)^2} dE_s,
\end{aligned} \tag{2.61}
$$

with the number of SEs in the surface calculated above, the yield of SEs is

$$
\delta = N_s P_s = \frac{N(E_{pe}) P_s \lambda}{4R(E_{pe})} \left(1 - \exp\left(-1.6\frac{R(E_{pe})}{\lambda}\right)\right) \tag{2.62}
$$

when E_{pe} is far greater than E_F, $N(E_{pe})$ can be approximated as

$$
N(E_{pe}) = \frac{E_{pe}}{E_v}, \tag{2.63}
$$

then

$$
\delta = N_s P_s = \frac{E_{pe} P_s \lambda}{4E_v R(E_{pe})} \cdot \left(1 - \exp\left(-1.6\frac{R(E_{pe})}{\lambda}\right)\right). \tag{2.64}
$$

The SEY varies the PE energy. Assuming that the PE energy at the maximum SEY is E_m, and the electron range corresponding to the PE energy is R_m, then

$$
\left.\frac{\partial \delta}{\partial E_{pe}}\right|_{E_{pe} = E_m} = 0, \tag{2.65}
$$

The R_m calculated above satisfy the following equation

$$1.6\frac{R_m}{\lambda}=\left(1-\frac{1}{\alpha}\right)\left(\exp\left(1.6\frac{R_m}{\lambda}\right)-1\right),\qquad(2.66)$$

Let $R_m = R'\lambda$, R' satisfy the following equation

$$1.6R'=\left(1-\frac{1}{\alpha}\right)\left(\exp(1.6R')-1\right),\qquad(2.67)$$

when α is a certain value, R' can be obtained according to the above formula. Table 2.3 shows the R' values of different materials.

Getting the value of R', where $R_m = R(E_m)=\dfrac{L_k}{\rho_r}\left(\dfrac{E_m}{E_k}\right)^{\alpha}$, E_m can be obtained by

$$E_m = E_k\left(\frac{R'\rho_r\lambda}{L_k}\right)^{1/\alpha},\qquad(2.68)$$

It is easy to get from the above

$$\frac{R\left(E_{pe}\right)}{R_m}=\left(\frac{E_{pe}}{E_m}\right)^{\alpha},\qquad(2.69)$$

so

$$\frac{R\left(E_{pe}\right)}{\lambda}=\left(\frac{E_{pe}}{E_m}\right)^{\alpha}\cdot\frac{R_m}{\lambda}=R'\left(\frac{E_{pe}}{E_m}\right)^{\alpha}.\qquad(2.70)$$

Substituting it into the above formula, then

TABLE 2.3   The R' Values of Different Materials

Metal	Atomic Number (Z)	R'
Al	13	1.19
Cu	29	1.39
Ag	47	1.53
Au	79	1.61

$$\delta = \frac{\delta_m}{1-\exp(-1.6R')} \left(\frac{E_{pe}}{E_m}\right)^{1-\alpha} \left(1-\exp\left(-1.6R'\left(\frac{E_{pe}}{E_m}\right)^{\alpha}\right)\right) \quad (2.71)$$

where

$$\delta_m = \frac{E_m P_s}{4 E_v R'} \left(1-\exp(-1.6R')\right). \quad (2.72)$$

2.3.1.3.3 Incident Angle   Many studies have shown that the SEY will increase significantly with the incident angle. When normal incidence, the range of PE is $R(E_{pe})$, and when the electron is incident into the material at an angle θ with the normal to the metal surface, the incident depth is $R\cos\theta$. Therefore, the distribution function of electrons at the depth direction should be revised as

$$n(z,\theta_{in}) = \frac{N(E_{pe})}{R(E_{pe})\cos\theta}, \quad (2.73)$$

Then, the contribution of the internal SEs in the range of $z \sim z + dz$ to the surface SEs is

$$n_s(z)dz = \frac{N(E_{pe})}{2R(E_{pe})\cos\theta} \left(\exp\left(-\frac{z}{\lambda}\right) - z\Gamma\left(0,\frac{z}{\lambda}\right)\right)dz. \quad (2.74)$$

The upper limit of the integral for calculating the number of SEs on the surface should be corrected as $R\cos\theta$.

$$N_s(\theta) = \int_0^{R(E_{pe})\cos\theta} n_s(z)dz$$

$$= \frac{N(E_{pe})}{2R(E_{pe})\cos\theta} \int_0^{R(E_{pe})\cos\theta} \left(\exp\left(-\frac{z}{\lambda}\right) - z\Gamma\left(0,\frac{z}{\lambda}\right)\right)dz$$

$$= \frac{N(E_{pe})\lambda}{4R(E_{pe})\cos\theta} F_\Gamma\left(\frac{R(E_{pe})\cos\theta}{\lambda}\right)$$

$$\approx \frac{N(E_{pe})\lambda}{4R(E_{pe})\cos\theta} \left(1-\exp\left(-1.6\frac{R(E_{pe})\cos\theta}{\lambda}\right)\right) \quad (2.75)$$

Let $\lambda_\theta = \lambda/\cos\theta_{in}$, only λ should be replaced by λ_θ in the formula of SEY when oblique incidence.

It is expressed by the maximum yield and its corresponding PE energy,

$$E_m(\theta_{in}) = \left(\frac{R'\rho\lambda_\theta}{B}\right)^{1/\alpha} = \left(\frac{R'\rho\lambda}{B\cos\theta_{in}}\right)^{1/\alpha} = E_m(0)\left(\frac{1}{\cos\theta_{in}}\right)^{1/\alpha} \quad (2.76)$$

The corresponding maximum yield is

$$\delta_m(\theta_{in}) = \delta_m\left(\frac{1}{\cos\theta_{in}}\right)^{1/\alpha} \quad (2.77)$$

Substitute it into the above formula,

$$\delta = \frac{\delta_m(\theta)}{1-\exp(-1.6R')}\left(\frac{E_{pe}}{E_m(\theta)}\right)^{1-\alpha}\left(1-\exp\left(-1.6R'\left(\frac{E_{pe}}{E_m(\theta)}\right)^{\alpha}\right)\right) \quad (2.78)$$

It is noteworthy that the trajectory of PEs is assumed to be a straight line along the incident direction when simulating the movement of electrons in the materials. For small incident angle, this assumption is true, but when the incident angle exceeds 60°, a large deviation will occur. Therefore, the above formula is applicable to the case when the incident angle is within 60°. In the actual multipactor process, electrons reciprocating move between the two metal walls and interact with the electromagnetic field to multiply. The oblique incident PEs will quickly leave the multipactor area due to the transverse movement, so they have little effect on multipactor. Therefore, the SEE characteristics with an incident angle within 60° are mainly considered when analysing the multipactor.

2.3.1.3.4 Backscattered Secondary Electrons   The BEs consist of two parts: the low-energy electron directly reflected by the surface barrier and the electron elastically scattered by the surface atom. The SE yield of low-energy backscattering directly reflected by the surface barrier is

$$\eta_1(E_{pe}) = \eta_1(0)\exp\left(-\frac{E_{pe}}{E'_{BS}}\right), \quad (2.79)$$

where $\eta_1(0)$ is a low-energy backscatter yield. For an ideal clean surface, its value is 0; for the material with adsorption on the surface, its value is about 0.6, and the E'_{BS} value is 12.5 eV.

Cazaux gives the electron yield reflected by the surface atom at normal incidence is

$$\eta_2\left(E_{pe},0\right)=\eta_2(\infty)\left(1-\exp\left(-\frac{E_{pe}}{E_{BS}}\right)\right) \tag{2.80}$$

$\eta_2(\infty)$ is the high-energy backscattering coefficient and depends on the material. The value of E_{BS} is also related to the material.

Furman developed a probability statistical model of SEY in 2002, where the backscatter electron yield varies with incident angle θ:

$$\eta_2\left(E_{pe},\theta\right)=\eta_2\left(E_{pe},0\right)\left(1+e_1\left(1-\cos{}^{e_2}\theta\right)\right), \tag{2.81}$$

where e_1 and e_2 are correction parameters. For commonly used metal materials, e_1 and e_2 are 0.8 and 0.4, respectively. Thus, the corrected backscatter yield for oblique incidence can be obtained.

The total elastic backscatter coefficient is

$$\eta=\eta_1+\eta_2 \tag{2.82}$$

Some PEs are elastically backscattered on the surface, cannot enter the material and cannot excite the internal SEs, so the PEs that contribute to TSEs account for $(1-\eta)$ of all PEs. The total electron emission yield is

$$\delta_{total}=\eta+(1-\eta)\delta \tag{2.83}$$

2.3.2 Monte Carlo Simulation

The scattering cross-section of electrons should first be calculated to analyse the scattering process. The scattering cross-sections of electrons are also divided into elastic scattering cross-sections and inelastic scattering cross-sections according to different scattering types. The elastic scattering cross-section of the PE is the integration of the differential elastic scattering cross-section of the PE in all directions, while the inelastic scattering cross-section of the PE is the integration of the differential inelastic scattering cross-section of the PE in all directions and various energy losses.

2.3.2.1 Elastic Scattering

In this chapter, the Mott scattering model is used to calculate the differential scattering cross-section. The Mott differential elastic scattering cross-section can be expressed as:

$$\frac{d\sigma_e}{d\Omega} = |f(\theta)|^2 + |g(\theta)|^2, \tag{2.84}$$

where σ_e is the elastic scattering cross-section ($cm^2 \cdot atom^{-1}$), Ω is the solid angle (deg). $f(\theta)$ and $g(\theta)$ are the incident and scattering wave division functions, which can be obtained by the wave division method. The total scattering cross-section of Mott elastic scattering can be obtained by integrating Mott differential scattering cross-section in all directions.

$$\sigma_e = 2\pi \int_0^\pi \frac{d\sigma_e}{d\Omega} \sin\theta \, d\theta \tag{2.85}$$

2.3.2.2 Inelastic Scattering

In addition, the inelastic scattering process with energy loss will occur between electron and material atom. In this chapter, the fast SE model is used to simulate the electron with energy higher than 3 keV, and the Penn dielectric function model is used when the energy is lower than 3 keV.

2.3.2.2.1 Fast Secondary Electron Model   In the simulation process of the fast SE model, the differential cross-section $d\sigma_{in}/d\Omega$ of the inelastic scattering between the electron and the sample is expressed as:

$$\frac{d\sigma_{in}}{d\Omega} = \frac{\pi e^4}{E^2}\left(\frac{1}{\Omega^2} + \frac{1}{(1-\Omega)^2} - \frac{1}{\Omega(1-\Omega)}\right), \tag{2.86}$$

where E is the electron energy(keV), Ω is the energy loss coefficient and τ is the ratio of the kinetic energy of the electron to the static mass energy.

The Bethe energy loss formula modified by Joy and Luo is used in the fast SE model to simulate the energy loss in the inelastic scattering process:

$$\left(\frac{dE}{dS}\right)_{Bethe} = 78,500\frac{\rho Z}{AE}\ln\left(\frac{1.166(E+kJ)}{J}\right) \tag{2.87}$$

The unit is $keV \cdot cm^{-1}$, where A is the atomic mass($g \cdot mol^{-1}$), Z is the atomic number of the material, J is the average ionization energy(keV), ρ is the density($g \cdot cm^{-3}$), and k is the correction coefficient of the ionization energy.

2.3.2.2.2 Penn Dielectric Function Model   In this chapter, the Penn dielectric function model is used when the electron energy is lower than 3 keV. The inelastic differential scattering cross-section of the collision between electron and atom is as follows:

$$\frac{d\sigma_{in}(E,\hbar\omega)}{d(\hbar\omega)} = \frac{me^2}{2\pi\hbar^2 NE}\text{Im}\left[-\frac{1}{\varepsilon(0,\hbar\omega)}\right]S\left(\frac{\hbar\omega}{E}\right), \qquad (2.88)$$

where E is the PE energy, m is the electron mass, N is the molecular number density of the material, e is the electron charge and $\hbar\omega$ is the energy loss.

In addition, the interaction between the electron and phonon and the interaction between the electron and polaron are also considered in the Penn model, so it is necessary to determine the inelastic scattering type in the simulation process. In this chapter, it is determined by the reciprocal of total mean free path of inelastic scattering:

$$\lambda_{in}^{-1} = \left(\lambda_{in\text{-}e}\right)^{-1} + \left(\lambda_{in\text{-}ph}\right)^{-1} + \left(\lambda_{in\text{-}po}\right)^{-1} \qquad (2.89)$$

With a random number $R_{in\text{-}Penn}$, which scattering process occurs is determined by calculating the probability of each scattering, i.e., $\lambda_{in}/\lambda_{in\text{-}e}$, $\lambda_{in}/\lambda_{in\text{-}ph}$ and $\lambda_{in}/\lambda_{in\text{-}po}$.

2.3.2.3 Simulation Process of Electron Scattering
For the scattering of a large number of electrons in the samples, the Monte Carlo method is used to simulate the scattering process. The simulation flow is shown in Figure 2.13:

1. Calculate the step of the electron according to the energy of the PE, and the coordinates of the next time position are obtained;

2. With a random number $R_{e\text{-}in}$ from 0 to 1, whether elastic scattering occurs is determined by the ratio of the reciprocal of elastic free path

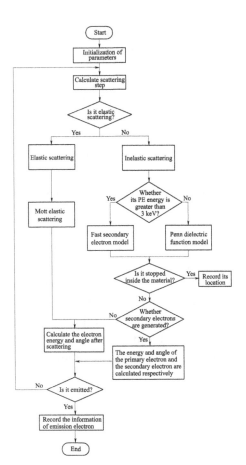

FIGURE 2.13   Monte Carlo simulation flow of electron scattering process.

λ_e^{-1} and the reciprocal of inelastic free path λ_{in}^{-1} to the reciprocal of the total mean free path $\lambda_m^{-1} = \lambda_e^{-1} + \lambda_{in}^{-1}$, i.e., λ_m/λ_e and λ_m/λ_{in};

3. When elastic scattering occurs, the angle of the scattered electron is calculated according to the Mott elastic scattering model. If inelastic scattering occurs, the inelastic scattering models should be chosen according to the energy of the PE first: the fast quadratic electron model should be used when it is larger than 3 keV, and the Penn dielectric function model should be used when it is smaller than 3 keV. In the fast SE model, when the energy lost by scattering is larger than the forbidden band width of dielectric material, a SE is generated and a hole is left at the collision position. In the Penn model,

when an electron is scattered by another electron in the inelastic scattering model, a SE will be produced and a hole will be left;

4. After coordinate transformation of the scattered electrons and the newly generated SEs, the next is to judge whether it will leave the sample or not. If this electron emit from the sample, finish its tracking and record the emission; if it still stays inside the sample, go to step 1 and continue to track its next scattering.

When all the electrons are tracked, the charge distribution of all the internal deposits, the number of emission electrons and the yield (SEY) are recorded.

Figure 2.14 shows the simulation and experiment data of SEY curve of PMMA material in the range of 100 eV~10 keV. Here, the simulation and experiment data is the SEY of sample without charge. Therefore, the simulation method above can be applied to both metal and dielectric materials without charge. In the experiments of this book, a new neutralization method is used to minimize the effect of dielectric charge. The simulation and experimental results in Figure 2.14 can also further verify the reliability of the model.

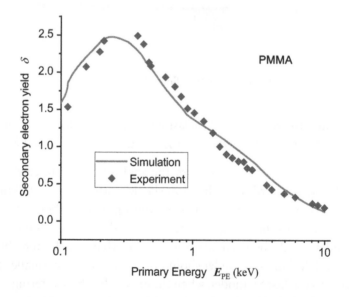

FIGURE 2.14 Comparison between simulation and experiment of SEY curve of PMMA materials.

2.4 MEASUREMENT OF SEY

Because the SEs are generated in the depth of several nanometers under the surface of the material, the energy of the SEs is low (only a few electron volts), so the SEY is very sensitive to the surface of the material. This section introduces several measurement methods and SE emission measurement equipment. The SEY and SEE spectra data of some common metal materials are finally summarized by analysing factors that affect the SEY of the material.

2.4.1 Measurement of SEY of Metal Materials

The measurement of the SEY of metal materials is relatively simple compared with that of the medium materials. At present, the most commonly used method is the electronic gun method. In this method, the electronic gun is used as a stable electronic source. The SEY can be obtained by hitting on the sample to be tested by focused electron beam and measuring the SE current collected [24]. Compared with the previous triode and four electrode tube methods, the electron gun method has significantly improved in the reliability and accuracy of the measurement. The electron gun method can also analyse the SEY of angular incidence.

Taking the measurement system of the SEY of metal developed by the key laboratory of space microwave of Xi'an Institute of space radio technology (Figure 2.15) as an example, the test method of the SEY of metal materials is introduced in detail [25,26]. The PE is supplied by two electron guns with 20 eV–5 keV and 3 keV–30 keV beam energy, respectively.

FIGURE 2.15   Measurement equipment of SEY of metal.

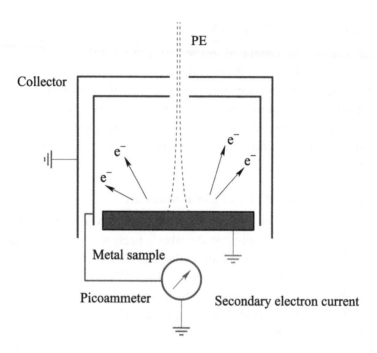

FIGURE 2.16   Collector test principle.

Two methods can be used to collect SEs: collector and bias DC methods. The measuring principle of the collector method is as shown in Figure 2.16: when the sample stage is connected with the collector, the current measured by picoammeter is the PE current, which is recorded as I_p; under the same conditions, when the sample is disconnected with collector, the current measured on the collector is the SE current, which is recorded as I_s.

$$SEY = \frac{SE \quad current}{PE \quad current} = \frac{I_s}{I_p} \tag{2.90}$$

The bias DC method measures the PE current and the SE current approximately by applying different bias voltage to the sample. As shown in Figure 2.17, when a large positive bias voltage is applied to the sample, the current measured by picoammeter is approximately the PE current I_p; the current measured when applying a negative bias voltage to the sample is the current difference I_r between the PE current and the SE current. With them, the SEY of the material can be obtained.

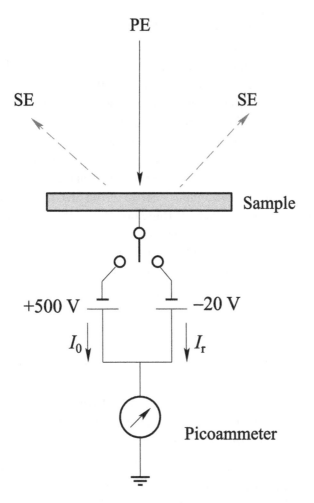

FIGURE 2.17   Test schematic diagram of SEY current method.

The bias current method and collector method have their own advantages. Theoretically, the collector is better for collecting SEs, but the operation of the collector method is not as convenient as that of the bias current method because the source electron beam must be aligned with the collector hole and the alignment of the sample and collector must be kept in the collector method. In addition, the size of the collector limits the moving of the sample. The bias current method is simple and fast, but the current of the source electron is smaller than the actual value because the high-energy electrons may escape from the sample surface.

2.4.2 Measurement of SEY of Dielectric and Semiconductor Materials

The SEE characteristics of dielectric and semiconductor materials are basically similar to those of metals, but the measurement is relatively difficult and complex due to their surface electrification. When these materials are irradiated by electron, the charge will be accumulated on their surface. The positive charges are accumulated at the second electron emission coefficient $\delta > 1$; and the negative charge are accumulated at the second electron emission coefficient $\delta < 1$. The charge accumulation changes the surface potential, the PE incident energy, the escape energy of the SE and the BE, leading to the continuous change of the SEY measured until the dynamic equilibrium is reached. Therefore, the method of measuring the SEY of metal is not suitable for that of dielectric materials.

For dielectric and semiconductor materials, the DC method and collector method can also be used under certain conditions [27,28]. However, in order to reduce surface electrification, there are generally two improvement measures as follows:

1. If the source electron beam with short pulse is used, the phenomenon of surface charge can be ignored at the shorter pulse duration and the lower repetition rate. This method is generally used for semiconductor materials. This method is not applicable for the material with poor conductivity because the generated surface charge cannot leak out within the pulse gap.

2. A low-energy electron gun is used to neutralize the surface charge and balance the surface potential of the material to be measured within the time initial electron beam pulse stops.

For example, the SEE measuring instrument developed by Utah State University is shown in Figure 2.18, which is equipped with two electron guns (with energy of 50 eV–5 keV and 4 keV–30 keV, respectively) to provide PEs and one low-energy electron gun (with energy of less than 1 eV) to neutralize the surface electrification of dielectric materials during the test.

In recent years, Belhaj et al. proposed a new method, Kelvin probe method, to measure the SEY of dielectric materials [29,30]. As shown in Figure 2.19, the energy of the two electron guns is 1 eV–2 keV and 1 keV–22 keV, respectively. The SE detection device consists of a Faraday cup, collection stage, electronic energy analyser and Kelvin probe, which can be calibrated by the collector method and Kelvin probe method.

The test process of the Kelvin probe method is shown in Figure 2.20. First, the surface potential of the dielectric material is measured by the Kelvin probe before the test, and then a negative bias voltage V_{is} is applied to ensure that there is no electron escape during the pulse test; the surface of the sample is irradiate by a single pulse with a charge amount of Q_i, and the stabilized surface potential V_{it} is measured again. The difference of the surface potential $\nabla V_i = (V_{it} - V_{is})$ is related to the SEY of material.

FIGURE 2.18   SEE measurement instrument developed by Utah State University.

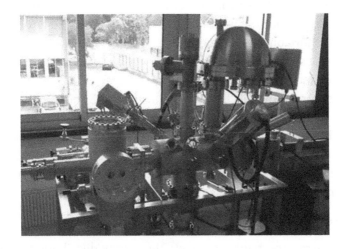

FIGURE 2.19   Experimental equipment of the French Aerospace Materials Research Agency.

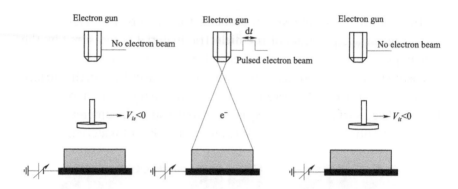

FIGURE 2.20   14 Kelvin probe test process.

The parasitic capacitance of the system is C. When δ is less than 1, negative charges accumulated on the sample surface; when δ is greater than 1, positive charges accumulated. Between two pulses, the sample surface is discharged, and the SEY is calculated by

$$\delta = 1 - \frac{C\Delta V_i}{Q_i} \tag{2.91}$$

Compared with the DC method or collector method, the Kelvin method is more suitable for the measurement of the SEY of dielectric materials. The parasitic capacitance is used for quantitative calculation to avoid errors caused by surface discharge. The advantages of high sensitivity and low noise reduce the system error to a certain extent and significantly improve the measurement accuracy of the SEY. Of course, the Kelvin probe method also has its shortcomings, including the complicated experimental process and time-consuming separate measurement and calculation of the parasitic capacitance of the system.

In summary, although the development of SEE measurement equipment has a long history, there are still many problems that need to be improved, such as the effect of the PE incident angle on measuring the SEY, in-situ surface characteristics analysis, in-situ processing, neutralization of surface charge and so on. Therefore, the research on measuring equipment is still in progress and has a long way to go.

2.4.3 SES Measurement

The SEE energy spectrum represents the energy distribution of the SEs emitted when the electrons with a certain energy impact on the surface

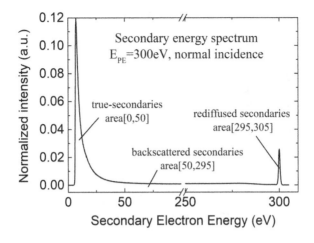

FIGURE 2.21 The SES of gold under 300 eV PE energy.

of the material. According to the generation processes of SEs, SEs include TSEs, BEs and inelastic scattered electrons. From the experimental point of view, these three types of SEs are distinguished according to the SE energy. Figure 2.21 is the SES curve of a gold material with a purity of 99.99% when the PE energy is 300 eV. Compared with SEY, which describes the probability of SEs emitted, the energy spectrum can describe the number and energy of SEs more intuitively and accurately.

As shown in Figure 2.21, the SES includes three parts, a TSE peak at lower energy, a backscattering peak at higher energy and an inelastic scattering part between the two peaks. The BE peak at E = E_{pe} is directly related to the PE energy. The position of the BE peak changes with the PE energy because these electrons are PEs directly reflected from the surface of the material; the TSE peaks appear in the range of 0~50 eV, in which the peak is near 0~10 eV, account for the vast majority of the total emitted electrons; there is no obvious peak in the inelastic scattering area, Spangenberg has summarized the following rules by summarizing the SEE characteristics data extensively:

1. TSEs account for about 90% of the emitted electrons;

2. Inelastic scattering electrons account for about 7% of the emitted electrons between 50 eV and 98% of the PE energy;

3. BEs are not emitted strictly but reflected PEs, which account for about 3% of the emitted electrons, and the energy is very close to the PE energy (usually 99% of the PE energy).

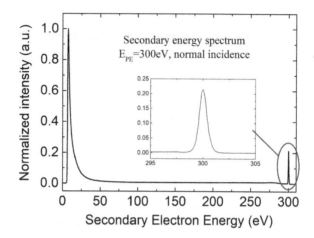

FIGURE 2.22   The maximum value-normalized SES.

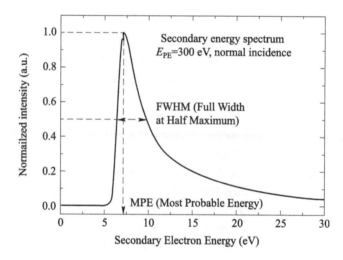

FIGURE 2.23   The maximum value-normalized TSE energy spectrum.

TSEs account for the majority of the SEs, so the TSE peaks should be focused on. The experimental SES corresponds to different SEYs. Therefore, the SES must be normalized before analysis. The common processing method is area normalization (Figure 2.21) and the maximum normalization (Figure 2.22). The maximum normalized SES mainly focuses on the two main parameters: FWHM and MPE, as shown in Figure 2.23. The FWHM is the peak width when the TSE peak falls to half, reflecting the energy distribution range and concentration area of the TSE; the MPE represents the energy of the SE corresponding to the peak, reflecting the energy corresponding to the largest number of TSEs. The area-normalized

SES is mainly used to represent the probability of SEs with different emission energies when electrons with a certain energy are incident. Therefore, the area-normalized SES is widely used in numerical simulation.

2.5 FACTORS AFFECTING SEY

The atomic number, crystal structure and activity of the material will cause the SEY to be very different. So far, the theoretical study of the effect of the material itself on the SEY has not been determined. Since SEs are generated within a few nanometers under the surface of the material, the energy of the SEs is low (only a few electron volts). Therefore, the SEY is extremely sensitive to the surface of the material, and any change of the surface morphology will vary with the SEY of the material.

2.5.1 Surface Adsorption and Contaminants on the Surface

In general, there will be a certain amount of adsorbed gases and contaminants on the surface of samples exposed to the atmosphere and some oxide layer on metals, which will vary with the SEY.

Electron will be affected by surface adsorption and contaminants when enter and exit from the surface. When the electrons enter the surface, the escape depth λ of the internal SEs will be different due to the adsorption and contaminants on the surface, thus affecting the incident energy E_m corresponding to δ_m; in addition, the surface potential barrier U and the expected value of the internal SE energy E_v alter, thus δ_m will be different. Assuming that the surface of the sample is saturated for a long time exposed to the atmosphere, the changes of the escape depth λ of the SEs are shown in Table 2.4.

The effect of surface adsorption on SEY can be studied by the heating desorption experiment. The silver-plated sample is heated to different temperatures and real-time monitor the pressure in the chamber. The variation of gas desorption amount Q_s with the sample temperature T_s is obtained, as shown in Figure 2.24.

TABLE 2.4   Changes of Electron Mean Free Path in Metals Exposed to Atmosphere

Materials	Atomic Number (Z)	Mean Free Path of Electron (λ nm^{-1})	Mean Free Path of Electron after Exposure to Atmosphere (λ nm^{-1})
Au	79	1.6	1.4
Ag	47	1.4	0.7
Cu	29	1.2	0.9
Al	13	3.8	3.7

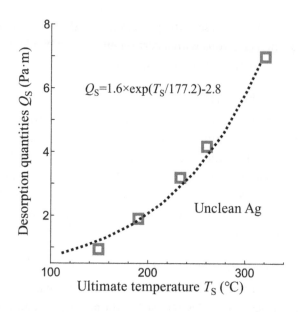

FIGURE 2.24 The desorption amount per unit area varies with the temperature of silver-plated sample.

If the gas is desorbed from the surface, the work function of the material surface changes, which affects the SEY. For example, the silver-plated sample is easy to adsorb gas in the atmosphere, so the gas will desorbed after heating, so that the surface work function increases and the SEY decreases. Figure 2.25 measured the SEY after thermal desorption of different temperatures. Figure 2.26 analyses the variation of SEY maximum value σ_m and work function ϕ with heating temperature T_s and the desorption amount Q_s

When gas adsorbed onto metals in the atmosphere, the water vapour, dust, carbon hydride, chloride, sulphide and fluoride will cover in the form of membrane or particle on the shallow layer, and the oxide layer will grow on the metal surface with exposure time. These containments will vary with the SEY of metal materials, the influence on SEY is analysed through the experiment of removing containments on the surface by Ar ion sputtering (Figure 2.27).

The surface potential barrier of uncleaned sample is close to the work function of the oxide because of the existence of oxide layer on the surface. The containments adsorbed on the sample surface will reduce the barrier and increase the SEY. Among the metals, dense Al_2O_3 passivation layer is easy to form on the surface of aluminium, which is difficult to remove by argon ion sputtering; the gold-plated material is stable and not easy to be oxidized, so the surface potential barrier changes little after argon ion sputtering.

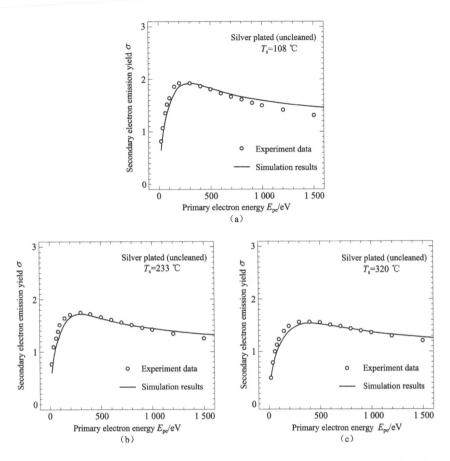

FIGURE 2.25 (a–c) SEY comparison of prediction and experimental results, the SEY of Ag heated at 108°C.

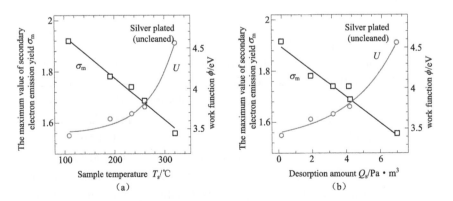

FIGURE 2.26 (a and b) Variation of the SEY maximum value σ_m and work function ϕ with heating temperature T_s and the desorption amount Q_s.

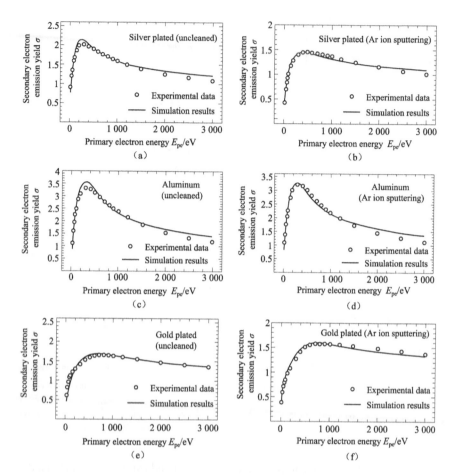

FIGURE 2.27   (a–f) Effect of Ar ion sputtering cleaning on the SEY of metal materials.

2.5.2 Surface Topography

When it comes to fluctuation on the metal surface, the SEs that can be emitted will impact on the side wall, which will change the trajectory and thus the electrons to be scattered or absorbed. Therefore, the SEY is very sensitive to the surface morphology of the material. Assuming that the cylindrical porous array shown in Figure 2.28 is fabricated on the silver-plated surface, and the parameters of the holes are as follows: H – hole depth, R – cylinder radius, D – distance between adjacent holes. Figure 2.28 illustrates the principle of SEY reducing of a porous surface. When electrons move in a single hole, some of them may be reflected or

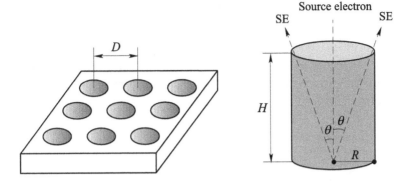

FIGURE 2.28 A model of a porous array fabricated on a silver-plated surface. H – hole depth, R – cylinder radius, D – distance between two adjacent holes.

absorbed by the side wall when impacting. Finally, only a small amount of electrons can escape from the hole and emit from the surface.

Assuming that a certain amount of electrons enter the hole with a given energy and direction, the emitted SEs are tracked by numerical simulation to determine the electrons will escape, absorb, or impact the wall to generate new SEs. Assuming that these electrons move linearly in vacuum, the SEY of the porous silver-plated surface σ_{porous} is calculated by weighted averaging the SEY of the plane and the hole.

$$\sigma_{\text{porous}} = P\sigma_{\text{hole}} + (1 - P)\sigma_{\text{flat}} \tag{2.92}$$

where σ_{flat} is the SEY of silver plating material, σ_{hole} is the SEY of a single hole and the average, P is the distribution probability of holes, which is defined as the ratio of porous area to the whole surface.

$$P = \frac{\text{Porous area}}{\text{Whole surface}} = \frac{\pi R^2}{D^2} \tag{2.93}$$

The aspect radio is defined as

$$A_R = \frac{H}{2R} \tag{2.94}$$

The SEY of silver-plated materials, $\sigma_{\text{flat}} = 2.5$, is obtained by experimental measurement and the SEY of a single hole is as

$$\sigma_{\text{hole}} = \sigma_{\text{bottom}} \cdot P_{\text{escape}} \tag{2.95}$$

where σ_{bottom} is the SEY at the bottom in the hole, P_{escape} is the probability of the emitted SEs escaping out of the hole. For the calculation of the SEY of the porous surface, assume that the SEY at the bottom $\sigma_{bottom} = 2.5$ is the same as that of the silver-plated surface. As shown in Figure 2.28, if an electron hits the centre of the bottom, its maximum polar angle allowed for electron emission is as

$$\theta = \arctan\left(\frac{R}{H}\right) \tag{2.96}$$

The escape probability of the electron is

$$P_{escape} = \int_0^\theta \cos(\varphi)\,d\varphi = \sin(\theta) \tag{2.97}$$

Substituting equations (2.93)–(2.97) into equation (2.92), the SEY of the silvered porous surface can be calculated. In Figure 2.29, the SEY of the porous surface varies with the aspect radio and porosity is analysed. It can be seen that increasing the aspect radio and porosity will reduce the SEY, However, the trend that increasing aspect ratio decreases SEY is a non-linear "diminishing return" behaviour. Thus, pursuing higher and higher aspect ratios to reduce SEY is a misguided objective. Design of the porous

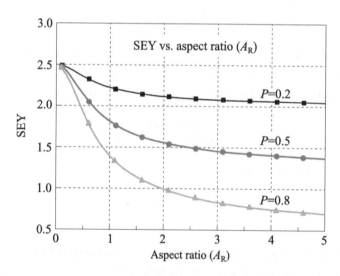

FIGURE 2.29 Relationship between the SEY and the aspect ratio and porosity of porous surface $\sigma_{flat} = \sigma_{bottom} = 2.5$.

surface invites our consideration of fabrication difficulty and insert loss that affect the device's ability to propagate RF waves. Equation (2.92) can be rearranged in form ($y = ax + b$),

$$\sigma_{porous} = P(\sigma_{hole} - \sigma_{flat}) + \sigma_{flat} \qquad (2.98)$$

It can be seen that the inverse relation of SEY with the porosity exists only when $\sigma_{hole} < \sigma_{flat}$.

The porous structure fabricated on the silver-plated surface by three nano-manufacturing techniques of sputtering, photolithography and etching, respectively, and the SEY changes of the silver-plated porous surface are measured [31]. Prolonged ion bombardment on silver-plated materials may cause the materials to be eroded. Figure 2.30a is an SEM photograph of the silver-plated surface after sputtering. The gap depth is about 219 nm, and the SEY measurement results show that the SEs emission coefficient can be dropped from 2.5 to 1.8 by 1.5 keV argon ion sputtering of the silver-plated sample. There are two reasons for the reduction of the SEY. One is to remove the surface containments, and the other is the groove formed on the surface that changes both the incident angle of the PE and the escape probability of the SE. In Figure 2.30b, the porous array pattern on the photomask is engraved on the silver-plated substrate by photolithography, the obtained parameters of the porous silver-plated surface by the laser scanning microscope (LSM) are as follows, the average pore depth of 4.9 μm, average pore diameter of 13.7 μm, porosity of 0.57, and aspect ratio of 0.36, the maximum value of the measured SEY is 1.47. In Figure 2.30c, irregular silver-plated porous surface manufactured by wet chemical etching, in which Ag is oxidized as $AgNO_3$ in the mixed solution HNO_3 and HF, HF is to promote the reaction rate and accelerate the etching process. Many microcracks are formed at the grain boundaries through the etching reaction. The measurement result of the maximum value of the SEY is 1.2, reduced by 52%.

In fact, these nanofabrication techniques can also be used on other metals to reduce the SEY. Three-dimensional microhole arrays can reduce the SEY more effectively than two-dimensional grooves.

2.6 THE SEY AND SES OF SOME COMMON METAL MATERIALS

The SEY and SES of the metal materials summarized in this chapter are the experimental results from the equipment introduced in Section 2.3.1.

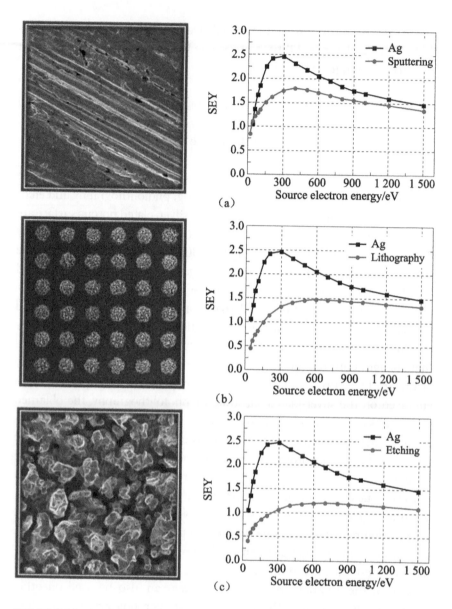

FIGURE 2.30   (a–c) SEY of the silver-plated porous structure and its surface.

As the input parameters for multipactor simulation analysis, the range of 40–1000 eV of SEY and the energy spectrum of true SEs in the range of 0–50 eV are given (Figure 2.31).

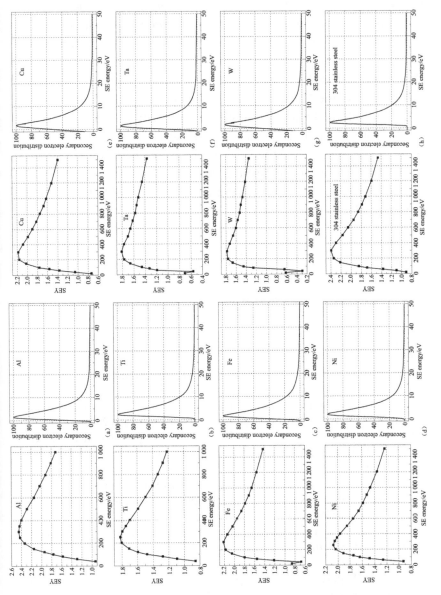

FIGURE 2.31 (a–h) Typical electron emission coefficient and energy spectrum of aerospace materials.

2.7 SUMMARY

This chapter mainly describes the basic theory, numerical simulation methods, measuring techniques and some experimental results of SEE. The SEE is expounded from the whole physical processes including collision, excitation, movement and emission of electrons in the material. For the numerical simulation of the SEE process, two methods are introduced: the Monte Carlo-based particle simulation method and empirical theory model (including the Furman model, Everhart model and semi-physics model). Through the typical SEE test equipment, this chapter introduces two most commonly used methods to measure the SEY, the collector method and the bias DC method, as well as the methods to improve the surface electrification of the medium, analyses the factors affecting the SEY of the material, and summarizes the SEY and the SES data of some common metal materials.

REFERENCES

1. Austin L, Starke H. The reflection of cathode rays and a new occurrence of secondary emission connected with it. *Annals of Physics*, 1902, 9(10):271–292.
2. Zhao Y. *Electron Beam Ion Beam Technology.* Xi'an: Xi'an Jiaotong University Press, 2002
3. Seiler H. SEE in the scanning electron microscope[J]. *Journal of Applied Physics*, 1983, 54(11):R1–R18.
4. Reimer L. *Scanning Electron Microscopy: Physics of Image Formation and Microanalysis.* Berlin: Springer-Verlag, 1985.
5. Ding Z, Wu Z, Sun X et al. *Microanalytical Physics and Its Applications.* Anhui: China University of science and Technology Press, 2009.
6. Bishop H, Riviere J. Estimates of the efficiencies of production and detection of electron-excited Auger emission. *Journal of Applied Physics*, 1969, 40(4):1740–1744.
7. Powell CJ. Precision, accuracy, and uncertainty in quantitative surface analyses by Auger-electron spectroscopy and x-ray photoelectron spectroscopy. *Journal of Vacuum Science and Technology*, 1990, 8(2):735–763.
8. Cazaux J. SEE and charging mechanisms in auger electron spectroscopy and related e-beam techniques. *Journal of Electron Spectroscopy*, 2010, 176(1–3):58–79.
9. Hatch AJ. Suppression of multipacting in particle accelerators. *Nuclear Instruments and Methods*, 1966, 41(2):261–271.
10. Baglin V, Collins I, Henrist B, et al. A summary of main experimental results concerning the SEE of copper. Geneva, 2002.

11. Rosenberg RA, Mcdowell MW, Ma Q, et al. X-ray photoelectron spectroscopy and secondary electron yield analysis of Al and Cu samples exposed to an accelerator environment. *Journal of Vacuum Science and Technology*, 2003, 21(5):1625–1630.

12. Suetsugu Y, Shirai M, Shibata K, et al. Development of a bellows chamber with a comb-type RF shield for high-current accelerators. *Nuclear Instruments and Methods*, 2004, 531(3):367–374.

13. Padamsee H, Joshi A. SEE measurements on materials used for superconducting microwave cavities. *Journal of Applied Physics*, 1979, 50(2):1112–1115.

14. Zameroski ND, Kumar P, Watts C, et al. Secondary electron yield measurements from materials with application to collectors of high-power microwave devices. *IEEE Transactions on Plasma Science*, 2006, 34(3):642–651.

15. Cheng GX, Liu L. Effect of surface produced secondary electrons on the sheath structure induced by high-power microwave window breakdown. *Physics of Plasmas*, 2011, 18(3):033507.

16. Charbonnier F. Arcing and voltage breakdown in vacuum microelectronics microwave devices using field emitter arrays: Causes, possible solutions, and recent progress. *Journal of Vacuum Science and Technology*, 1998, 16(2):880–887.

17. Vicente C, Mattes A, Wolk D, et al. Contribution to the RF breakdown in microwave devices and its prediction. Washington, 2006: 22–27.

18. de Lara J, Perez F, Alfonseca M, et al. Multipactor prediction for on-board spacecraft RF equipment with the MEST software tool. *IEEE Transactions on Plasma Science*, 2006, 34(2):476–484.

19. Lai ST, Tautz M. On the anticritical temperature for spacecraft charging. *Journal of Geophysical Research*, 2008, 113(A11):A11211.

20. Balcon N, Payan D, Belhaj M, et al. SEE on space materials: Evaluation of the total secondary electron yield from surface potential measurements. *IEEE Transactions on Plasma Science*, 2012, 40(2):282–290.

21. Ferguson DC. New frontiers in spacecraft charging. *IEEE Transactions on Plasma Science*, 2012, 40(2):139–143.

22. Bruining H. *Physics and Applications of Secondary Electron Emission*. London: Pergamon Press, 1955: 93–96.

23. Wittry DB, Kyser DF. Cathodoluminescence at p-n Junctions in GaAs[J]. *Journal of Applied Physics*, 1965, 36(4):1387–1389.

24. Farnsworth HE. Electronic bombardment of metal surfaces. Physical Review, 1925, 25:41.

25. Zhang N, Cao M, Cui WZ, Zhang HB. Lab-Built Test-Stand for SEE of Metals in Ultra-High Vacuum. ISSN:1672-7126.2014.05.

26. Cui W, Yang J, Zhang N. Testing method of the secondary electron emission yield of space metal materials. *Space Electronic Technology*, 2013, 10(2):75–78.

27. Boubaya M, Blaise G. Charging regime of PMMA studied by SEE. *European Physical Journal-Applied Physics*, 2007, 37(1):79–86.

28. Blaise G, Pesty F, Garoche P. The SEE yield of muscovite mica: Charging kinetics and current density effects. *Journal of Applied Physics*, 2009, 105(3):034101.

29. Belhaj M, Tondu T, Inguimbert V, Chardon JP. A Kelvin probe based method for measuring the electron emission yield of insulators and insulated conductors subjected to electron irradiation. *Journal of Physics D: Applied Physics*, May 2009, 42(10):105–309.

30. Belhaj M, Paulmier T, Guibert N, et al. New secondary electron and irradiation facilities at ONERA: a bridge between eV and MeV energy range. *MULCOPIM*, 2014, 11:331–336

31. Yang J, Cui W, Li Y, et al. Investigation of argon ion sputtering on the SEE from gold samples. *Applied Surface Science*, 2016, 382:88–92.

CHAPTER 3

Electromagnetic Particle-in-Cell Method

3.1 OVERVIEW

As an important numerical simulation method for studying the beam-wave interaction process, the particle-in-cell (PIC) method has been widely used in many research fields, such as the controlled thermonuclear fusion, space physics and astrophysics, plasma physics and vacuum electronics. With the further improvement in the performance of high-speed and large-capacity computers, it is bound to further promote the development of PIC method research and its application field, expand the scope of research and application, shorten the research and application cycle and promote the development of some emerging disciplines. The PIC model can be divided into many types according to different standards, for example, the electrostatic model, magnetostatic model and electromagnetic model are classified according to electromagnetic field processing; one-dimensional (1D), two-dimensional (2D) (including 2.5-dimensional) and three-dimensional (3D) models to the calculation space and velocity dimension; the bounded and unbounded models to calculation boundary; relativistic and non-relativistic situations to the particle energy; in addition, there are point particles, finite-size particles, and weighted particle models and so on.

In addition to the development of the dimensional model from one dimension to three dimensions, the most common model is classified according to the three major types of electrostatic, magnetostatic and electromagnetic model. These three models have a lot in common in basic theory; the main difference is only the emphasis on the calculation and application of electromagnetic fields. The electrostatic model is used to simulate the physical problem that electrostatic force plays a decisive role. In the simulation, it is not necessary to solve the complex Maxwell equations, but it is enough to solve the Poisson equation. The magnetostatic model is used to simulate only the physical problems of low-frequency self-consistent magnetic fields, such as the Alfvén wave, pinch, and ion cyclotron in plasma. In the simulation, the displacement current term in the Maxwell equations is removed, and it is not necessary to use the complete Maxwell equations. The electromagnetic model is used to simulate physical problems related to electromagnetic radiation or self-consistent electromagnetic fields. In the simulation, it is not necessary to solve the complex Maxwell equations. Because of the completeness of the electromagnetic model, electrostatic model and magnetostatic model can be regarded as the simplified form of the electromagnetic model under certain conditions, so electrostatic problem and magnetostatic problem can also be solved by the electromagnetic model, but the calculation speed will be slower. The PIC method corresponding to the electromagnetic model is called the electromagnetic particle-in-cell (EM-PIC) method, which is the most widely used particle simulation method. The EM-PIC method will be introduced in this book for the electromagnetic model that is essential for the study of the beam-wave interaction involved in the phenomenon of secondary electron multiplication.

EM-PIC is a first-principle method based on Maxwell equations under the consideration of the beam-wave interaction in vacuum, which is widely used in vacuum electronic device research and design. In practice, it can improve the device performance, reduce the research and development cycle and experimental expenses. However, in the EM-PIC method, the complete Maxwell equations, Lorentz force equations and particle motion equations must be solved, and this method is also limited by the time stability conditions, so the calculation amount is relatively large, especially in the 3D program. It takes days or even weeks for the calculation of a slightly complicated vacuum electronics. The parallel algorithm is currently the best way to solve this problem, and the Message Passing Interface (MPI) model is the first choice for the parallel algorithm, which is mostly used in excellent PIC software [4].

From a methodological perspective, the EM-PIC method is a combination of the finite difference time domain (FDTD) method and the PIC method. The basic idea of the FDTD method is to solve discrete Maxwell equations in discrete grids with discrete time steps. In the PIC method, the movement of the particles is propelled in a discrete grid under the consideration of the interaction between the field and the particles. The motion process of particles in the electromagnetic field can be obtained through the mutual influence and promotion.

3.2 DEVELOPMENT AND APPLICATION OF EM-PIC METHOD

The PIC method was originally specifically aimed at plasma simulation research. Since the 1950s, Buneman [1] and Dawson [2] began to simulate the thin layer of electrons and directly calculated the field from Coulomb's law and Gauss' law. It was only in the 1960s that the introduction of discrete grids enabled the PIC method to be truly realized. In that era, the PIC method was mainly used in a 1D periodic system, and the number of particles that can be simulated is also quite limited. This method can simulate some of the collective effects of plasma, especially the space charge effect, so it has been increasingly applied to various types of plasma problems, and many good results have been achieved. The work of Birdsall and Langdon [3], Hockney and Eastwood [4] in the 1970s and 1980s represented the continuous development of particle simulation and published some classic works. The emergence of space gridding and finite-size particle models has made PIC theory systematic and formalized.

Since the 1990s, 2D and 3D PIC methods have been basically finalized [5]. For ease of use, object-oriented ideas have been implemented and applied in particle simulation software [6]. It is also from this period that massively parallel computing began to take the stage of history [7–9]. Since this century, the PIC method has been basically mature, but many new ideas and methods [10–20] have been proposed. In the field of high-performance parallel computing, there are also many research results [21–23].

Since the PIC method has a very important application prospect in the field of national defence high-tech, a large number of well-known universities and research institutions are committed to this area. Since the 1970s, research on plasma particle simulation methods has also begun in China, and many achievements have been made in the past few decades. It is precisely because of the excellent application prospects of this method that many foreign excellent software has embargoed on China, so in the past decade, many universities and research institutions of China, such as the University of Electronic Science and Technology of China, National University of Defense Technology, Xi'an Jiaotong University, China Academy of Engineering Physics and Northwest Institute of Nuclear Technology, which represent the development level of China, have also begun to develop various types of particle simulation software.

After decades of development, there are many excellent EM-PIC software [24]; here we take a few typical excellent software as an example to give a brief introduction.

MAGIC [25], an old commercial EM-PIC software, was developed by MRC company since the 1970s. The software is implemented by the FDTD method combined with the particle-in-cell (FDTD-PIC) method. It is available for Windows, Linux and Unix operating systems. The development of parallel versions began in the 1990s, and it currently supports parallel computing of MPI. As a general and authoritative electromagnetic PIC code, it supports a variety of coordinate systems and a lot of common boundary conditions, can simulate many complex physical problems and get good and reliable results. The disadvantage is that the user interface is relatively simple, and it is prone to memory limitations for large-scale devices that are currently embargoed to China.

The electromagnetic PIC KARAT [26] developed by Dr. Vladimir P. Tarakanov of Russia and MAFIA [27] developed by the German CST (Computer Simulation Technique) company are also common PIC software in China. These two software are widely used in China and their performance is relatively good. However, there is still a certain gap comparing to MAGIC.

For excellent parallel electromagnetic particle simulation software, there is the 3D electromagnetic PIC software LSP [28] developed by the US MRC company and the 3D particle simulation software ICEPIC [29] developed by the US Air Force Laboratory. The full name of LSP is Large Scale Plasma, which is a software for large-scale plasma simulation. ICEPIC is short form for Improved Concurrent Electromagnetic Particle-in-Cell, which is also implemented by the FDTD-PIC method, mainly for high-performance parallel computing. Its parallel algorithm supports multi-dimensional partitioning and dynamic load balancing with high parallel computing efficiency, representing the highest level in the world today. The disadvantage is that it currently supports only rectangular coordinate systems; the implementation of boundary conditions is not quite perfect and does not support large-scale HPM device simulation.

In addition to the above representative software, there are many well-known PIC software, such as PIC3D and VIPER [30] in the United Kingdom, QUICKSILVER [31] in the United States and object-oriented OOPIC of the University of California. It is worth mentioning that China's research time in this area is still short, but it has developed very rapidly. Some particle simulation algorithms [32–35] and software [36–41] studies conducted by the Beijing Institute of Applied Physics and Computational Mathematics and the National University of Defense Technology can perform a good simulation of the interaction process of intense lasers and plasmas. For the general-purpose electromagnetic particle simulation software, several relatively mature general-purpose particle simulation software have also been developed, such as CHIPIC by the University of Electronic Science and Technology [42,43], UNIPIC jointly by the Northwest Institute of Nuclear Technology and Xi'an Jiaotong University [44,45] and NEPTUNE [46,47] by the Beijing Institute of Applied Physics and Computational Mathematics.

NEPTUNE was developed by the Beijing Institute of Applied Physics and Computational Mathematics in 2005. The software is based on the J Adaptive Structured Mesh Applications Infrastructure (JASMIN). With its main feature of large-scale parallel computing, it is currently a 3D version and can also simulate some high-power microwave source devices more accurately.

UNIPIC jointly developed by Xi'an Jiaotong University and Northwest Institute of Nuclear Technology since 2002 is based on the windows platform. The UNIPIC-2D introduced later can perform more accurate simulations of more high-power microwave source devices and has been certified by relevant departments. The development of a 3D version and a parallel version [48] began in 2005.

CHIPIC software was developed by the University of Electronic Science and Technology on the Window operating system from 2002. The 2D version was originally launched, and its credibility has been certified by relevant departments. The development of CHIPIC3D began in 2005, and the development of the parallel version [49,50] began at the same time. The parallel function has been successfully implemented. However, due to the short time of software development, there are still many areas that need to be improved (such as a wide variety of boundary conditions).

3.3 PROCEDURE OF THE EM-PIC METHOD

The general realization process of the EM-PIC method is that the initial electromagnetic field environment and charged particles (or a certain particle generation law) are given, and then the particles and field are gradually updated and propelled from the initial conditions, as shown in Figure 3.1, where ρ is the charge density, J is the current density, E and B are the electromagnetic field, v is the particle velocity, r is the particle displacement, and s is the distance from the particle position to the surface of the grid in which it is located.

Obviously, in each iteration, the field of the particle position in the continuous space is obtained according to the field value on the discrete grid, thereby propelling the particle motion, which in turn affects the source items (charge density and current) in the field equation density to update the field value, allowing the cycle to continue. In the calculation process, it is also necessary to deal with different field boundary conditions and particle boundary conditions. Considering collision, the Monte Carlo collision method should also be introduced.

Figure 3.2 shows a time-discrete model of the EM-PIC method (that is, the leapfrog model) [51–54]. Obviously, it not only includes the discrete model of the FDTD method but also contains the iteration of physical quantities related to the particle such as the particle force F, momentum p, and displacement x. In terms of the field update, the electric field E is calculated in integer time steps, and the magnetic field B is calculated in half time steps; in particle processing, the particle force F and displacement x are solved in integer time steps, and the particle momentum p is solved in half time step.

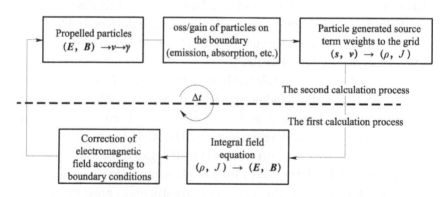

FIGURE 3.1   EM-PIC flow chart.

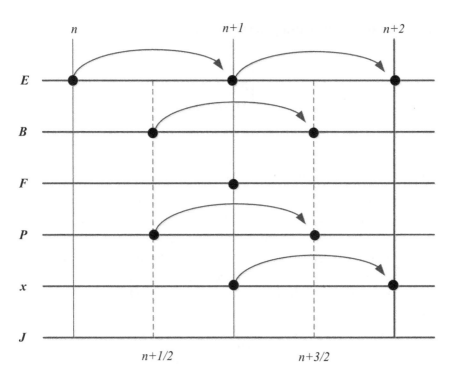

FIGURE 3.2   Schematic diagram of EM-PIC leapfrog.

3.4 FDTD METHOD

3.4.1 Maxwell Equations and Differential Difference Scheme

The update of the FDTD electromagnetic field is based on the Maxwell equations. The expression of Maxwell time domain differential equations [55] is:

$$\nabla \cdot \boldsymbol{D} = -\rho \qquad (3.1)$$

$$\nabla \times \boldsymbol{E} = -\frac{\partial \boldsymbol{B}}{\partial t} - \boldsymbol{M} \qquad (3.2)$$

$$\nabla \times \boldsymbol{H} = \frac{\partial \boldsymbol{D}}{\partial t} + \boldsymbol{J} \qquad (3.3)$$

$$\nabla \cdot \boldsymbol{B} = 0 \qquad (3.4)$$

The constitutive relationship in isotropic linear media is

$$\begin{cases} D = \varepsilon_0 E \\ B = \mu_0 H \\ M = \sigma_m H \\ J = \sigma E \end{cases} \tag{3.5}$$

where $\varepsilon_0 = 8.554 \times 10^{-12}\,\mathrm{F\,m^{-1}}$ is the vacuum dielectric constant, $\mu_0 = 4\pi \times 10^{-7}\,\mathrm{H\,m^{-1}}$ is the vacuum permeability, σ_m is the permeability with the unit $\Omega\,\mathrm{m^{-1}}$ and σ is the conductivity with the unit $\mathrm{S\,m^{-1}}$.

Since the divergence equation is satisfied during the FDTD update process, only the curl equation needs to be considered to derive the FDTD difference format. Taking the 3D rectangular coordinate system as an example, equations (3.2) and (3.3) can be changed into the following equations:

$$\frac{\partial H_z}{\partial y} - \frac{\partial H_y}{\partial z} = \frac{\partial D_x}{\partial t} + \sigma E_x \tag{3.6}$$

$$\frac{\partial H_x}{\partial z} - \frac{\partial H_z}{\partial x} = \frac{\partial D_y}{\partial t} + \sigma E_y \tag{3.7}$$

$$\frac{\partial H_y}{\partial x} - \frac{\partial H_x}{\partial y} = \frac{\partial D_z}{\partial t} + \sigma E_z \tag{3.8}$$

$$\frac{\partial E_z}{\partial y} - \frac{\partial E_y}{\partial z} = -\frac{\partial B_x}{\partial t} - \sigma_m H_x \tag{3.9}$$

$$\frac{\partial E_x}{\partial z} - \frac{\partial E_z}{\partial x} = -\frac{\partial B_y}{\partial t} - \sigma_m H_y \tag{3.10}$$

$$\frac{\partial E_y}{\partial x} - \frac{\partial E_x}{\partial y} = -\frac{\partial B_z}{\partial t} - \sigma_m H_z \tag{3.11}$$

In the FDTD algorithm, the geometric space of the problem is discretized into spatial grid points, on which the electric field and magnetic field components are assigned, which then solves the Maxwell equations in a discrete time manner. From formula (3.6) to formula (3.11), in the FDTD calculation, the space and time derivatives in Maxwell equations are approximated with the finite difference method, by which a set of equations are constructed, and then the instantaneous field of the next step is calculated on the basis of that of the previous time step, thereby constructing an algorithm of time forward to simulate the time domain process of the electromagnetic field. First, here the discrete method of the system of equations is discussed; then the discrete space, discrete time model and the method of constructing discrete equations will be elaborated one by one.

Any continuous function can be sampled with discrete points. In theory, the original function can be well approximated as long as there are enough sampling points. The approximate accuracy depends on the sampling rate. Another factor that has a significant impact on accuracy is the choice of discrete operators. There usually are a variety of discrete operators, of which a differential operator is the focus of discussion here. Assuming that $f(x)$ is a continuous function of discrete point sampling, its derivative can be expressed as

$$f'(x) = \lim_{\Delta x \to 0} \frac{f(x + \Delta x) - f(x)}{\Delta x} \tag{3.12}$$

where Δx is a non-zero minimum constant, so the above formula can be approximated as

$$f'(x) \approx \frac{f(x + \Delta x) - f(x)}{\Delta x} \tag{3.13}$$

which is the difference format of the differential expression. Since the sampling point in the front is used here, it is also called the forward difference. Correspondingly, the format of backward difference and centre difference is

$$f'(x) \approx \frac{f(x) - f(x - \Delta x)}{\Delta x} \tag{3.14}$$

$$f'(x) \approx \frac{f(x + \Delta x) - f(x - \Delta x)}{2\Delta x} \tag{3.15}$$

The centre difference format can be regarded as the average of the forward difference and the backward difference, which theoretically has

TABLE 3.1   The Difference Formats of the First and Second Derivatives of the Function

	First Derivative $\partial f / \partial x$		Second Derivative $\partial^2 f / \partial x^2$	
	Differential Format	Accuracy	Differential Format	Accuracy
Forward difference	$\dfrac{f_{i+1} - f_i}{\Delta x}$	First-order accuracy	$\dfrac{f_{i+2} - 2f_{i+1} + f_i}{(\Delta x)^2}$	First-order accuracy
Backward difference	$\dfrac{f_i - f_{i-1}}{\Delta x}$	First-order accuracy	$\dfrac{f_i - 2f_{i-1} + f_{i-2}}{(\Delta x)^2}$	First-order accuracy
Central difference	$\dfrac{f_{i+1} - f_{i-1}}{2\Delta x}$	Second-order accuracy	$\dfrac{f_{i+1} - 2f_i + f_{i-1}}{(\Delta x)^2}$	Second-order accuracy

higher accuracy. Table 3.1 summarizes the different difference formats and related accuracy of common first and second derivatives.

In general, the centre difference is the most common difference format, with high accuracy and the most widely used. However, in many cases, the difference of first-order precision also has enough calculation accuracy, and the forward or backward difference has better simplicity in the calculation format and so on. The differential formats with higher accuracy are less used due to the complex calculation format and the intuitiveness, which is not conducive to programming.

3.4.2 Spatial Discrete and Time-Discrete Format

The so-called Yee cell, the core of FDTD, was first proposed by Yee. He proposed that the sampling nodes of E and H field components should be arranged alternately in space and time. Each E (or H) field component is surrounded by four H (or E) field components. With this discrete method, Maxwell equations can be transformed into difference equations, and the spatial electromagnetic field can be solved step by step on the time axis. This kind of discretization can be divided into spatial and temporal discretization. The spatial discretization model is generally called Yee grid model, and the temporal discretization model is generally called frog model.

Figure 3.3a shows a Yee grid model [1,43–48] in the rectangular coordinate system, wherein the electric and magnetic field components in three directions are, respectively, represented by $E1$, $E2$, $E3$ and $B1$, $B2$ and $B3$. Obviously, each magnetic field component is surrounded by four electric field components; similarly, each electric field component is surrounded by four magnetic field components. Not only the spatial sampling method of

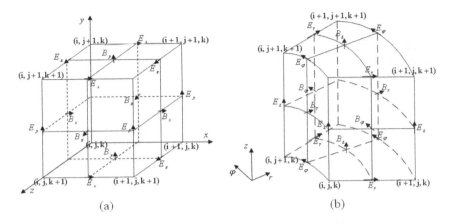

FIGURE 3.3   (a and b) Three-dimensional Yee grid element in two coordinate systems.

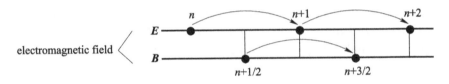

FIGURE 3.4   Frog format of electromagnetic field calculation.

electromagnetic field conforms to the natural structure of Faraday induction law and Ampere law but also the spatial relative position of each component of electromagnetic field is suitable for the difference calculation of the Maxwell equation, which can properly describe the propagation characteristics of the electromagnetic field. It can also be seen from the figure that the components of electric field and magnetic field in different directions are located in the positions of full grid or half grid in different directions in the traditional central difference method, which is with second-order accuracy. The Yee grid model in cylindrical coordinates in Figure 3.3b shows that its electromagnetic field components are basically similar to those in the rectangular coordinate system.

As shown in Figure 3.4, the electric field is solved in integral time steps and the magnetic field is solved in half integral time steps. This discretization and spatial discretization form the basic discretization scheme of

TABLE 3.2   Node Positions of E and H Components in the Yee Cell

Electromagnetic Field Component		Spatial Component Sampling			Time Axis Sampling
		X Direction	Y Direction	Z Direction	
Node E	E_x	$i+1/2$	j	k	n
	E_y	i	$j+1/2$	k	
	E_z	i	j	$k+1/2$	
Node H	H_x	i	$j+1/2$	$k+1/2$	$n+1/2$
	H_y	$i+1/2$	j	$k+1/2$	
	H_z	$i+1/2$	$j+1/2$	k	

the FDTD method. Obviously, this scheme and the particle parameter impelling together constitute the complete discrete scheme of EM-PIC, as shown in Figure 3.2.

It can be seen from Table 3.2 that variables can be divided into full time step variables and half time step variables according to the time point of variable calculation. These variables are sampled and updated alternately in time sequence. The time interval between sampling and updating is half a time step away from each other, which ensures only sampling and updating once in each time step and can realize fast iterative solution in time. Table 3.2 shows the integer and semi-integer conventions of the spatial nodes and time steps of E and H components in the Yee cell. As listed in Table 3.2, rectangular coordinate is taken as an example. For cylindrical coordinates or spherical coordinates, the component format is unchanged. It is only necessary to change x, y, z corresponding to r, φ, z or r, θ, φ.

Assumed that $f(x,y,z,t)$ is the component in a certain direction of E or H, whose symbol in the discrete-time and spatial domains is represented as

$$f(x,y,z,t) = f^n(i,j,k) \tag{3.16}$$

where i,j,k is the grid index number, which is $x = i\Delta x, y = j\Delta y, z = k\Delta z$ under the uniform grid. For the convenience of discussion, the following discussion assumes that the grid is uniform without special description. Based on the node location in Table 3.2, the approximate expression of the time and space difference of the first partial derivative is

$$\begin{cases}\left.\dfrac{\partial f(x,y,z,t)}{\partial x}\right|_{x=i\Delta x} \approx \dfrac{f^n\left(i+\dfrac{1}{2},j,k\right)-f^n\left(i-\dfrac{1}{2},j,k\right)}{\Delta x}\\[2em]\left.\dfrac{\partial f(x,y,z,t)}{\partial y}\right|_{y=j\Delta y} \approx \dfrac{f^n\left(i,j+\dfrac{1}{2},k\right)-f^n\left(i,j-\dfrac{1}{2},k\right)}{\Delta y}\\[2em]\left.\dfrac{\partial f(x,y,z,t)}{\partial z}\right|_{z=k\Delta z} \approx \dfrac{f^n\left(i,j,k+\dfrac{1}{2}\right)-f^n\left(i,j,k-\dfrac{1}{2}\right)}{\Delta z}\\[2em]\left.\dfrac{\partial f(x,y,z,t)}{\partial t}\right|_{t=n\Delta t} \approx \dfrac{f^{n+1/2}(i,j,k)-f^{n-1/2}(i,j,k)}{\Delta t}\end{cases} \quad (3.17)$$

3.4.3 Difference Scheme

According to the difference scheme described in the previous section, the difference formula of the corresponding coordinate system can be obtained by expanding Maxwell equations in different coordinate systems. For the sake of simple discussion, the rectangular coordinate system will be taken as an example in this book, the difference formula in other coordinate systems can be similarly derived.

In rectangular coordinates, based on equation (3.17), the difference of equation (3.6) can be expanded to obtain

$$\frac{\left(D_x\right)^{n+1}_{i+1/2,j,k}-\left(D_x\right)^{n}_{i+1/2,j,k}}{\Delta t}+\sigma(i+1/2,j,k)\left(E_x\right)^{n+1/2}_{i+1/2,j,k}$$
$$=\frac{1}{\Delta y}\left[\left(H_z\right)^{n+1/2}_{i+1/2,j+1/2,k}-\left(H_z\right)^{n+1/2}_{i+1/2,j-1/2,k}\right]-\frac{1}{\Delta z}\left[\left(H_y\right)^{n+1/2}_{i+1/2,j,k+1/2}-\left(H_y\right)^{n+1/2}_{i+1/2,j,k-1/2}\right]$$

$$(3.18)$$

where

$$\left(D_x\right)^{n+1}_{i+1/2,j,k}=\varepsilon(i+1/2,j,k)\left(E_x\right)^{n+1}_{i+1/2,j,k} \quad (3.19)$$

$$\left(E_x\right)^{n+1/2}_{i+1/2,j,k}=\frac{\left(E_x\right)^{n+1}_{i+1/2,j,k}+\left(E_x\right)^{n}_{i+1/2,j,k}}{2} \quad (3.20)$$

For the sake of simplicity of the formula, the grid index of the conductivity and dielectric constant is no longer pointed out below, and their indices is consistent with the electric field or magnetic field component after it. Substituting formula (3.19) and formula (3.20) into formula (3.18), we can get

$$
\left(E_x\right)_{i+1/2,j,k}^{n+1} = \frac{2\varepsilon - \Delta t\sigma}{2\varepsilon + \Delta t\sigma}\left(E_x\right)_{i+1/2,j,k}^{n}
$$

$$
+ \frac{2\Delta t}{2\varepsilon + \Delta t\sigma}\left[\frac{\left(H_z\right)_{i+1/2,j+1/2,k}^{n+1/2} - \left(H_z\right)_{i+1/2,j-1/2,k}^{n+1/2}}{\Delta y} - \frac{\left(H_y\right)_{i+1/2,j,k+1/2}^{n+1/2} - \left(H_y\right)_{i+1/2,j,k-1/2}^{n+1/2}}{\Delta z}\right]
$$

$$(3.21)$$

Correspondingly, the 3D FDTD formula in the rectangular coordinates of formula (3.7) to formula (3.11) can be expressed as:

$$
\left(E_y\right)_{i,j+1/2,k}^{n+1} = \frac{2\varepsilon - \Delta t\sigma}{2\varepsilon + \Delta t\sigma}\left(E_y\right)_{i,j+1/2,k}^{n}
$$

$$
- \frac{2\Delta t}{2\varepsilon + \Delta t\sigma}\left[\frac{\left(H_z\right)_{i+1/2,j+1/2,k}^{n+1/2} - \left(H_z\right)_{i-1/2,j+1/2,k}^{n+1/2}}{\Delta x} - \frac{\left(H_x\right)_{i+1/2,j+1/2,k}^{n+1/2} - \left(H_x\right)_{i-1/2,j+1/2,k}^{n+1/2}}{\Delta z}\right]
$$

$$(3.22)$$

$$
\left(E_z\right)_{i,j,k+1/2}^{n+1} = \frac{2\varepsilon - \Delta t\sigma}{2\varepsilon + \Delta t\sigma}\left(E_z\right)_{i,j,k+1/2}^{n}
$$

$$
- \frac{2\Delta t}{2\varepsilon + \Delta t\sigma}\left[\frac{\left(H_x\right)_{i,j+1/2,k+1/2}^{n+1/2} - \left(H_x\right)_{i,j-1/2,k+1/2}^{n+1/2}}{\Delta y} - \frac{\left(H_y\right)_{i+1/2,j,k+1/2}^{n+1/2} - \left(H_y\right)_{i-1/2,j,k+1/2}^{n+1/2}}{\Delta x}\right]
$$

$$(3.23)$$

$$
\left(H_x\right)_{i,j+1/2,k+1/2}^{n+3/2} = \frac{2\mu - \Delta t\sigma_m}{2\mu + \Delta t\sigma_m}\left(H_x\right)_{i,j+1/2,k+1/2}^{n+1/2}
$$

$$
+ \frac{2\Delta t}{2\mu + \Delta t\sigma_m}\left[\frac{\left(E_y\right)_{i,j+1/2,k+1}^{n+1} - \left(E_y\right)_{i,j+1/2,k}^{n+1}}{\Delta z} - \frac{\left(E_z\right)_{i,j+1,k+1/2}^{n+1} - \left(E_z\right)_{i,j,k+1/2}^{n+1}}{\Delta y}\right]
$$

$$(3.24)$$

$$\left(H_y\right)^{n+3/2}_{i+1/2,j,k+1/2} = \frac{2\mu - \Delta t \sigma_m}{2\mu + \Delta t \sigma_m}\left(H_y\right)^{n+1/2}_{i+1/2,j,k+1/2}$$

$$-\frac{2\Delta t}{2\mu + \Delta t \sigma_m}\left[\frac{\left(E_x\right)^{n+1}_{i+1/2,j,k+1} - \left(E_x\right)^{n+1}_{i+1/2,j,k}}{\Delta z} - \frac{\left(E_z\right)^{n+1}_{i+1,j,k+1/2} - \left(E_z\right)^{n+1}_{i,j,k+1/2}}{\Delta x}\right]$$

$$(3.25)$$

$$\left(H_z\right)^{n+3/2}_{i+1/2,j+1/2,k} = \frac{2\mu - \Delta t \sigma_m}{2\mu + \Delta t \sigma_m}\left(H_z\right)^{n+1/2}_{i+1/2,j+1/2,k}$$

$$-\frac{2\Delta t}{2\mu + \Delta t \sigma_m}\left[\frac{\left(E_y\right)^{n+1}_{i+1,j+1/2,k} - \left(E_y\right)^{n+1}_{i,j+1/2,k}}{\Delta x} - \frac{\left(E_x\right)^{n+1}_{i+1/2,j+1,k} - \left(E_x\right)^{n+1}_{i+1/2,j,k}}{\Delta y}\right]$$

$$(3.26)$$

where Δx, Δy and Δz are the grid lengths in three directions, Δt is the time step, the superscript of the field component is the corresponding time step number, and the subscript is the corresponding grid index. Formula (3.21) to (3.26) is the standard difference scheme of the electromagnetic field in the rectangular coordinate system corresponding to Yee grid, also known as the centre difference scheme [53–55].

Careful readers can see that the FDTD formulas in the previous sections all consider the conductivity, permeability, dielectric coefficient and magnetic conductivity in the grid, so it is applicable to the calculation of general conductor, medium, vacuum and other components and belongs to the more general formula in the field of FDTD. However, in many calculations in the field of vacuum, especially in the calculation process of multipactor which needs to consider the influence of particles, in order to greatly reduce the amount of calculation, it is not necessary to comprehensively and elaborately consider the parameters such as conductor conductivity and permeability but only the materials such as vacuum, perfect conductor and isotropic dielectric. Only consider the electromagnetic fields in vacuum and dielectric, the corresponding FDTD calculation formula can be greatly simplified. Taking rectangular coordinate as an example, it can be simplified as follows:

$$\left(D_x\right)_{i+1/2,j,k}^{n+1} = \left(D_x\right)_{i+1/2,j,k}^{n} + \frac{\Delta t}{\Delta y}\left(H_z\right)_{i+1/2,j+1/2,k}^{n+1/2}$$

$$-\frac{\Delta t}{\Delta y}\left(H_z\right)_{i+1/2,j-1/2,k}^{n+1/2} + \frac{\Delta t}{\Delta z}\left(H_y\right)_{i+1/2,j,k-1/2}^{n+1/2} - \frac{\Delta t}{\Delta z}\left(H_y\right)_{i+1/2,j,k+1/2}^{n+1/2} \quad (3.27)$$

$$\left(B_x\right)_{i,j+1/2,k+1/2}^{n+3/2} = \left(B_x\right)_{i,j+1/2,k+1/2}^{n+1/2} + \frac{\Delta t}{\Delta z}\left(E_y\right)_{i,j+1/2,k+1}^{n+1}$$

$$-\frac{\Delta t}{\Delta z}\left(E_y\right)_{i,j+1/2,k}^{n+1} + \frac{\Delta t}{\Delta y}\left(E_z\right)_{i,j,k+1/2}^{n+1} - \frac{\Delta t}{\Delta y}\left(E_z\right)_{i,j+1,k+1/2}^{n+1} \quad (3.28)$$

$$\left(D_y\right)_{i,j+1/2,k}^{n+1} = \left(D_y\right)_{i,j+1/2,k}^{n} + \frac{\Delta t}{\Delta z}\left(H_x\right)_{i,j+1/2,k+1/2}^{n+1/2}$$

$$-\frac{\Delta t}{\Delta z}\left(H_x\right)_{i,j+1/2,k-1/2}^{n+1/2} + \frac{\Delta t}{\Delta x}\left(H_z\right)_{i-1/2,j+1/2,k}^{n+1/2} - \frac{\Delta t}{\Delta x}\left(H_z\right)_{i+1/2,j+1/2,k}^{n+1/2} \quad (3.29)$$

$$\left(B_y\right)_{i+1/2,j,k+1/2}^{n+3/2} = \left(B_y\right)_{i+1/2,j,k+1/2}^{n+1/2} + \frac{\Delta t}{\Delta x}\left(E_z\right)_{i,j+1/2,k+1}^{n+1}$$

$$-\frac{\Delta t}{\Delta x}\left(E_z\right)_{i,j,k+1/2}^{n+1} + \frac{\Delta t}{\Delta z}\left(E_x\right)_{i+1/2,j,k}^{n+1} - \frac{\Delta t}{\Delta z}\left(E_x\right)_{i+1/2,j,k+1}^{n+1} \quad (3.30)$$

$$\left(D_z\right)_{i,j+1/2,k}^{n+1} = \left(D_z\right)_{i,j+1/2,k}^{n} + \frac{\Delta t}{\Delta x}\left(H_y\right)_{i,j+1/2,k+1/2}^{n+1/2}$$

$$-\frac{\Delta t}{\Delta x}\left(H_y\right)_{i,j+1/2,k-1/2}^{n+1/2} + \frac{\Delta t}{\Delta y}\left(H_x\right)_{i-1/2,j+1/2,k}^{n+1/2} - \frac{\Delta t}{\Delta y}\left(H_x\right)_{i+1/2,j+1/2,k}^{n+1/2} \quad (3.31)$$

$$\left(B_z\right)_{i+1/2,j+1/2,k}^{n+3/2} = \left(B_y\right)_{i+1/2,j+1/2,k}^{n+1/2} + \frac{\Delta t}{\Delta y}\left(E_x\right)_{i+1/2,j+1,k}^{n+1}$$

$$-\frac{\Delta t}{\Delta y}\left(E_x\right)_{i+1/2,j,k}^{n+1} + \frac{\Delta t}{\Delta x}\left(E_y\right)_{i,j+1/2,k}^{n+1} - \frac{\Delta t}{\Delta x}\left(E_y\right)_{i+1,j+1/2,k}^{n+1} \quad (3.32)$$

Equation (3.27)–(3.32) is a simplified version of the rectangular coordinate difference scheme. In comparison, due to many abridged grid parameters, the storage and amount of calculation are greatly reduced, which is quite significant in the calculation of some complex computations. This difference scheme is not widely used in FDTD electromagnetic field calculation but in the calculation of microwave components in the vacuum environment, especially the beam-wave interaction. Due to the large demand of computing resources in the process of computing particles, some optimization measures are needed in the algorithm. Reducing the unnecessary grid parameters is one of the effective measures to reduce the amount of calculation.

3.5 PARTICLE MODEL AND EQUATION OF MOTION

3.5.1 Macroparticle Model

In practice, the total amount of particles in any device is more than the memory capacity of the general computer, so the use of simplified particle model is also an indispensable part of the work of EM-PIC method.

The implementation principle of macroparticle model is that a certain number of macroparticles are generated on the emission grid in each time step, each macroparticle represents a different number of real charged particles, and the charged amount of macroparticles is equal to the total charged amount of real particles [55]. By using the finite-size macroparticle model, the force between them can be well dealt with, so as to realize the collective characteristic [53].

It has been proved that the result is believable when enough macroparticles are used to represent real particles in the calculation area because the model has good statistical properties when the macroparticles reach a certain number.

3.5.2 Equations of Motion of Particles

Consider the interaction between particles and fields; the EM-PIC method is more similar to the actual situation, which is also an important factor known as the first principle. First, the influence of the field on particles, that is, the motion of particles in the electromagnetic field is considered. The equations of motion of particles in the relativistic case can be described by equations (3.33) and (3.34):

$$\frac{d}{dt}\gamma m v = q(E + v \times B) \tag{3.33}$$

$$\frac{d}{dt}x = v \qquad (3.34)$$

where x is the particle displacement, γ can be expressed by formula (3.35):

$$\gamma \equiv \frac{1}{\sqrt{1-(v/c)^2}} = \sqrt{1+(u/c)^2} \qquad (3.35)$$

$$u \equiv \gamma v \qquad (3.36)$$

The discrete method here is the frog scheme introduced earlier. According to the model in Figure 3.3, the above equations of motion can be changed into:

$$\frac{u^{n+1/2} - u^{n-1/2}}{\Delta t} = \frac{q}{m}\left(E^n + \frac{u^n}{\gamma^n} \times B^n\right) \qquad (3.37)$$

$$\frac{x^{n+1} - x^n}{\Delta t} = \frac{u^{n+1/2}}{\gamma^{n+1/2}} \qquad (3.38)$$

For the solution method of formula (3.37), Boris [56] gives a decomposition solution method, which conforms to the discrete property of EM-PIC method and is a more practical method:

$$u^- = u^{n-1/2} + \frac{q\Delta t E^n}{2m} \qquad (3.39)$$

$$u' = u^- + u^- \times \frac{q\Delta t B^n}{2\gamma^n m} \qquad (3.40)$$

$$u^+ = u^- + u' \times \frac{q\Delta t B^n}{m\gamma^n\left[1+\left(\Omega_c \Delta t/2\right)\right]^2} \qquad (3.41)$$

$$u^{n+1/2} = u^+ + \frac{q\Delta t E^n}{2m} \qquad (3.42)$$

It is not difficult to see that the particle motion is decomposed into two electric field actions and one magnetic field actions in this method, thus simplifying the complex solution process. Here, equations (3.39) and (3.42) correct the magnitude of momentum, and equations (3.40) and (3.41) correct the direction of momentum where $\Omega_c \Delta t \ll 1$ must be satisfied, wherein $\Omega_c = (q\mathbf{B}^n)/\gamma^n m$ means relativistic cyclotron frequency. $\mathbf{B}^n$ can be approximated as $(\mathbf{B}^{n-1/2} + \mathbf{B}^{n+1/2})/2$, γ^n can be calculated with $\mathbf{u}^- : \gamma^n = \sqrt{1 + (u^- / c)^2}$.

3.6 AN ALGORITHM OF BEAM-WAVE INTERACTION

According to the flow chart of EM-PIC method shown in Figure 3.1, electromagnetic field calculation and **beam-wave** interaction are the most basic and important algorithms in the EM-PIC method. Based on the particle model and motion equation in the previous section, this section focuses on the process and algorithm reality of **beam-wave** interaction. In order to facilitate the description, the rectangular coordinate system is also taken as an example for discussion and analysis. The case of the cylindrical coordinate system and polar coordinate system can be similarly derived.

3.6.1 Charged Particle Motion in Electromagnetic Fields

As can be seen from the Yee grid model in Figure 3.3, the electric field component is at the edge centre, while the magnetic field component is at the face centre, and the particles may exist at any position in the grid, so the equation (3.33) cannot be used directly. In order to achieve this goal, it can be divided into two steps: first, the weighted electromagnetic field components on the half grid are assigned to the whole grid, and then, the particle location is assigned by interpolation.

In the first step, the weighted electromagnetic field on the half grid is allocated to the whole grid. Consider that the average weighted electric field are assigned to the two ends of the grid line, and the average weighted magnetic field are assigned to the four corners of the grid face. Taking the field component in direction x of the whole grid point (i, j, k) as an example, there are the following weighting methods:

$$E_x(i,j,k) = \left(E_x(i-1/2,j,k) + E_x(i+1/2,j,k)\right)/2 \qquad (3.43)$$

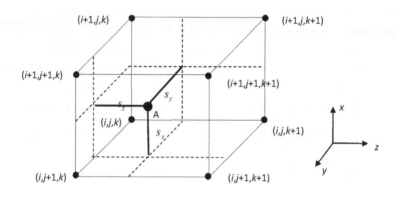

FIGURE 3.5   Schematic diagram of particle position in Yee grid.

$$B_x(i,j,k) = \Big(B_x(i,j-1/2,k-1/2) + B_x(i,j-1/2,k+1/2 +$$
$$\Big(B_x(i,j+1/2,k-1/2) + B_x(i,j+1/2,k+1/2) \Big)/4 \Big) \quad (3.44)$$

The weighting formula of electromagnetic field in other directions can be obtained similarly.

In the second step, the electromagnetic field on the whole grid is interpolated at the actual position of particles in the grid. As shown in Figure 3.5, it is assumed that the distance between particle A and the three coordinate surfaces is s_x, s_y and s_z, respectively. It is advisable to assume that

$$\begin{cases} \lambda_x = s_x/\Delta x \\ \lambda_y = s_y/\Delta y \\ \lambda_z = s_z/\Delta z \end{cases} \quad (3.45)$$

Then, the corresponding weighting factor w has the following formula:

$$w_{i,j,k} = \big(1-\lambda_x\big)\big(1-\lambda_y\big)\big(1-\lambda_z\big) \quad (3.46)$$

$$w_{i+1,j,k} = \lambda_x\big(1-\lambda_y\big)\big(1-\lambda_z\big) \quad (3.47)$$

$$w_{i,j+1,k} = \big(1-\lambda_x\big)\lambda_y\big(1-\lambda_z\big) \quad (3.48)$$

$$w_{i,j,k+1} = \left(1-\lambda_x\right)\left(1-\lambda_y\right)\lambda_z \tag{3.49}$$

$$w_{i+1,j+1,k} = \lambda_x\lambda_y\left(1-\lambda_z\right) \tag{3.50}$$

$$w_{i+1,j,k+1} = \lambda_x\left(1-\lambda_y\right)\lambda_z \tag{3.51}$$

$$w_{i,j+1,k+1} = \left(1-\lambda_x\right)\lambda_y\lambda_z \tag{3.52}$$

$$w_{i+1,j+1,k+1} = \lambda_x\lambda_y\lambda_z \tag{3.53}$$

where the subscript of w is the corresponding grid index. It can be concluded that the x direction field components at position of particle A are:

$$
\begin{aligned}
\left(E_x\right)_A &= w_{i,j,k}\left(E_x\right)_{i,j,k} + w_{i+1,j,k}\left(E_x\right)_{i+1,j,k} + w_{i,j+1,k}\left(E_x\right)_{i,j+1,k} \\
&+ w_{i,j,k+1}\left(E_x\right)_{i,j,k+1} + w_{i+1,j+1,k}\left(E_x\right)_{i+1,j+1,k} + w_{i+1,j,k+1}\left(E_x\right)_{i+1,j,k+1} \\
&+ w_{i,j+1,k+1}\left(E_x\right)_{i,j+1,k+1} + w_{i+1,j+1,k+1}\left(E_x\right)_{i+1,j+1,k+1}
\end{aligned} \tag{3.54}
$$

$$
\begin{aligned}
\left(H_x\right)_A &= w_{i,j,k}\left(H_x\right)_{i,j,k} + w_{i+1,j,k}\left(H_x\right)_{i+1,j,k} + w_{i,j+1,k}\left(H_x\right)_{i,j+1,k} \\
&+ w_{i,j,k+1}\left(H_x\right)_{i,j,k+1} + w_{i+1,j+1,k}\left(H_x\right)_{i+1,j+1,k} + w_{i+1,j,k+1}\left(H_x\right)_{i+1,j,k+1} \\
&+ w_{i,j+1,k+1}\left(H_x\right)_{i,j+1,k+1} + w_{i+1,j+1,k+1}\left(H_x\right)_{i+1,j+1,k+1}
\end{aligned} \tag{3.55}
$$

Similarly, the field components in other directions can be derived.

So far, the electromagnetic field component corresponding to each macroparticle can be clearly obtained, and then, the displacement and momentum change of the corresponding macroparticle can be obtained by formula (3.33) and (3.34).

3.6.2 Effect of Charged Particle Motion on Electromagnetic Field

The action of the **electromagnetic** field on the particles is a process of pushing the particles, i.e. changing their momentum and position. In this process, the space charge density and current density will inevitably be changed.

According to Maxwell equations, this change will also have a greater impact on the field. The following is a detailed analysis of this effect.

3.6.2.1 Effect of Particle Propulsion on Charge Density

Referring to the weighting field in the previous section and considering the effect on the space charge density, it is assumed that a particle is pushed to A in the grid of Figure 3.5, the weighted charge [3,4] can be expressed similarly as:

$$(q')_{i,j,k} = (1 - \lambda_x)(1 - \lambda_y)(1 - \lambda_z)q \tag{3.56}$$

$$(q')_{i+1,j,k} = \lambda_x(1 - \lambda_y)(1 - \lambda_z)q \tag{3.57}$$

$$(q')_{i,j+1,k} = (1 - \lambda_x)\lambda_y(1 - \lambda_z)q \tag{3.58}$$

$$(q')_{i,j,k+1} = (1 - \lambda_x)(1 - \lambda_y)\lambda_z q \tag{3.59}$$

$$(q')_{i+1,j+1,k} = \lambda_x\lambda_y(1 - \lambda_z)q \tag{3.60}$$

$$(q')_{i+1,j,k+1} = \lambda_x(1 - \lambda_y)\lambda_z q \tag{3.61}$$

$$(q')_{i,j+1,k+1} = (1 - \lambda_x)\lambda_y\lambda_z q \tag{3.62}$$

$$(q')_{i+1,j+1,k+1} = \lambda_x\lambda_y\lambda_z q \tag{3.63}$$

where q' is the weighted charges of the eight grid points occupied by the macroparticle, the subscript is the index of the corresponding grid point and q is the quantity of electric charge carried by the macroparticle.

In this way, the charge density of the entire grid point can be recalculated, and the calculation formula is:

$$\rho = \frac{Q}{\Delta x \Delta y \Delta z} \tag{3.64}$$

where Q is the sum of the charge weights of all macro-particles to the entire grid point.

3.6.2.2 Effect of Particle Propulsion on Current Density

Considering the influence of particle position on the spatial current density and assuming that the particle moves from point A to point B, from time n to time $n+1$, as shown in Figure 3.6, and its trajectory is parallel to the coordinate axis, there are a total of six random paths, as shown in Figure 3.7, that are available to allocate spatial current density. Under the known stability conditions, the macroparticles can only cross one grid at most in each time step, and this case will be described here as an example. Assuming that $r_x(i,j,k)$, $r_y(i,j,k)$ and $r_z(i,j,k)$ are the coordinates of the grid (i,j,k) in the x, y, and z directions, respectively.

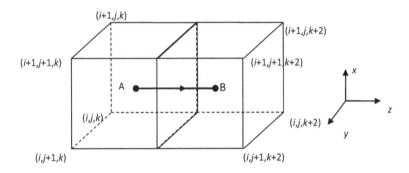

FIGURE 3.6 Schematic diagram of particle position change.

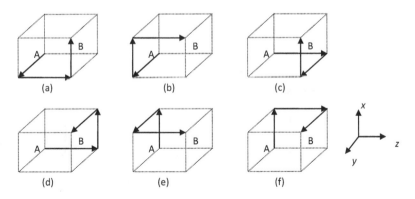

FIGURE 3.7 (a–f) Schematic diagram of random motion trajectories of particles.

Taking the particle moving from point A along the x direction to the plane yz where point B is located as an example, the motion is decomposed into two parts to form the current density. The current generated by the particle moving in the x direction, that is, the first part of the current density generated by the particle moving from the time n to the yz plane grid $(i+1, j, k)$ is:

$$J_{x1} = \frac{q_A \left(r_x(i+1, j, k) - r_{Ax}{}^n \right)}{\Delta x \Delta y \Delta z \Delta t} \tag{3.65}$$

The second part of the current density generated from the movement of the yz plane grid $(i+1, j, k)$ to the time $n+1$ (that is, the position of point B) is:

$$J_{x2} = \frac{q_A \left(r_{Ax}{}^{n+1} - r_x(i+1, j, k) \right)}{\Delta x \Delta y \Delta z \Delta t} \tag{3.66}$$

Then, the weighted current J_{x1} and J_{x2} are, respectively, assigned to the projection of the grid $(i+1, j, k)$ on the plane yz and are shown in Figure 3.8, and the position of the projection point is $(r_{Ay}{}^n, r_{Az}{}^n)$, s_y and s_z are the distances from the particle to the grid surface.

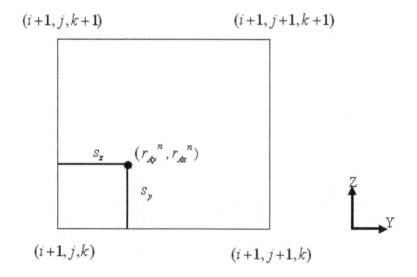

FIGURE 3.8   YoZ plane projection of current density distribution.

First, similar to the previous weighting method, it can be assumed that:

$$w_y = s_y \big/ \Delta y \ w_z = s_z \big/ \Delta z \tag{3.67}$$

giving J_{x1} the distribution expression as follows (J_{x2} can be obtained similarly):

$$
\left\{
\begin{aligned}
\Delta J_x(i,j,k) &= \left(1 - w_y\right)\left(1 - w_z\right) J_{x1} \\
\Delta J_x(i,j+1,k) &= w_y\left(1 - w_z\right) J_{x1} \\
\Delta J_x(i,j,k+1) &= \left(1 - w_y\right) w_z J_{x1} \\
\Delta J_x(i,j+1,k+1) &= w_y w_z J_{x1}
\end{aligned}
\right.
\tag{3.68}
$$

Similarly, the current density expression of particles moving in other directions can be obtained [4]. The effect of particle motion on the **electromagnetic** field can be obtained by substituting the above expression into the iteration formula.

3.7 PARTICLE BOUNDARY CONDITIONS

3.7.1 General Particle Boundary Conditions

Algorithmically, the charged particles are generally directly annihilated, when they move to the surface of a conductor or medium. However, sometimes consider the impact of secondary emission whose implementation process is relatively complicated, many documents [3,4,52,53] have studied the principles of this process.

In addition to the particle emission boundary condition, the perfectly matched layer (PML) is usually used on most of the boundary conditions mentioned in the previous part of the field boundary conditions. In the process of simulation, when charged particles move to these boundary surfaces, they are usually simply annihilated and moved out of the simulation area. Next, the particle processing on various boundary conditions is briefly analysed.

For conventional boundary conditions, except for the symmetrical boundary, on which the speed and position of particles must be reset according to the symmetry, on other boundaries such as conductors and media, particles are directly annihilated on the surface without the consideration of the very complex physical problems such as particle surface adsorption because these are beyond the scope of simulation.

For the boundary condition of the waveguide excitation source, when the charged particles move to the boundary, it is equivalent to have moved to the edge of the simulation area, so it is also directly annihilated; for the boundary condition of the equivalent excitation source, no additional treatment will be carried out for the particles because this kind of the boundary condition is placed inside the simulation area.

For electromagnetic wave absorbing boundary condition because it is equivalent to external open space, when charged particles move to the boundary surface, it is also equivalent to moving to the edge of simulation area, so the same annihilation treatment is carried out.

For other field boundary conditions, lumped parameter circuit components are placed inside the simulation area, and the size can be ignored. Generally, particles will not be processed on the boundary; for other boundary conditions placed on the edge of the simulation area, particles will be annihilated.

In addition to the above, there is a special kind of particle boundary conditions, such as metal grid (metal foil) and material sheet structure. Because these structures are widely used in some devices in recent years, it is necessary to discuss them. In the field treatment, the boundary conditions of this kind of structure is calculated as that of metal and medium, but in the particle treatment, consider the transmittance of particles, only part of particles are allowed to pass through. Take the metal grid (metal foil) as an example. When electrons hit the metal grid (metal foil) under the acceleration of the electric field, some of them are absorbed, some of them are scattered and the other part is transmitted through. The metal grid (metal foil) plays the role of shielding the field but not completely blocking particles.

3.7.2 Particle Emission Boundary Conditions

The particle emission boundary condition is an indispensable boundary condition in the EM-PIC method, which provides a way to inject macroparticles into the model. The different particle emission boundary conditions have the different physical mechanism, so the corresponding implementation algorithm and processing skills are also different. In order to simulate some common devices in the field of vacuum electronics, some commonly used models, such as hot electron emission, field emission, hot field emission, and space charge limited current emission, are discussed in detail in this section. As for the second electron emission, which has been discussed in the previous chapters, there will be no more discussion.

3.7.2.1 Thermal Electron Emission Boundary Conditions

Since Edison first observed the hot electron emission in vacuum in 1883, the hot cathode has been widely used in the traditional microwave vacuum electronics. This kind of cathode can provide a stable space charge limited current in the high vacuum environment for a long time, especially for low current or medium current.

For this kind of emission, the physical mechanism of electron emission should first be understood. Generally, this phenomenon can be explained with the free electron model, which can be expressed as that many electrons move freely in the metal in the thermal equilibrium state; the velocity distribution of the free electrons is subject to the Fermi-Dirac distribution; there is a potential barrier high enough on the metal surface, if the electrons want to escape from the metal, they must do a certain amount of work to overcome this barrier. The work function is the minimum energy that must be given by the electron with the maximum energy to overcome the potential barrier at a temperature of absolute zero.

Therefore, it is known that the thermal electron emission is the electron emission formed by using the heating method to increase the kinetic energy of the electrons inside the solid, so that some of them can escape out of the solid by overcoming the surface barrier.

The thermal electron emission of metals is generally described by the Richardson-Dushman formula [57–59]:

$$j = A_0 T^2 \exp\left[-\frac{\phi}{kT}\right] \qquad (3.69)$$

where $A_0 = \dfrac{4\pi emk^2}{h^3} = 1.204 \times 10^6 \left(\text{A} \cdot \text{m}^{-2} \cdot \text{K}^{-2}\right)$ is the theoretical value of the emission constant and the same for all metals. k is the Boltzmann constant, and φ is the work function of the metal.

The free electrons in metals may have a distribution of various states, but these states do not necessarily occupy electrons. When establishing the theory of solid electron emission, the number of electrons dn_E must also be known in the unit volume corresponding to the energy E to $E + dE$ interval. If the probability of electrons occupying the state of energy E is the Fermi-Dirac distribution function $f(E)$, and hence

$$dn_E = f(E)\frac{4\pi(2m)^{3/2}E^{1/2}}{h^3}dE \qquad (3.70)$$

where

$$f(E) = \frac{1}{\exp\left[\dfrac{E - E_F}{kT}\right] + 1} \qquad (3.71)$$

where k is the Boltzmann constant, T is the absolute temperature and E_F is the Fermi level.

It can be seen from equation (3.71) that when T is zero, the probability of electrons occupying all energy levels lower than E_F is 1, and the probability of energy levels higher than E_F is 0. The electrons in the metal at the Fermi level have the greatest energy. As the temperature increases, a few electrons with energy slightly lower than E_F occupy energy levels higher than E_F. The higher the temperature, the more electrons occupy higher energy levels than E_F.

The thermal electron emission often works under the acceleration field, and the work function will be reduced according to the Schottky effect. It can be described by the following formula:

$$j = j(0)e^{\frac{0.439\sqrt{E}}{T}} \qquad (3.72)$$

where $j(0)$ is the thermal electron emission current density at zero field and E is the electric field intensity on the cathode surface. This formula can generally be used to qualitatively represent the thermoelectron emission density under the acceleration field and can also be used to represent the temperature-limiting flow on the cathode surface.

When the electric field on the cathode surface is close to zero, equation (3.72) degenerates into equation (3.69), which means that equation (3.69) is still applicable at zero field. However, field emission will occur when the electric field strength is large to a certain extent (generally above 10^8 V m^{-1}).

The influence of field emission must be considered. This will be discussed in detail in the next section.

When the space charge density formed by the emitted electrons is sufficiently large, the initial velocity of the thermal electrons cannot be regarded as zero. The influence of the initial velocity and its distribution must be considered. In general, the initial velocity of thermally emitted electrons is considered to satisfy the Maxwell-Boltzmann distribution:

$$\frac{d^6 n}{d^3 v d^3 x} = \exp\left(-\frac{mv^2}{2kT}\right) \tag{3.73}$$

$$\frac{d^6 j_{\text{emit}}}{dv d\phi d\theta d^3 x} = v^3 \sin(\theta)\cos(\theta)\exp\left(-\frac{mv^2}{2kT}\right) \tag{3.74}$$

Logarithm is taken for both sides of formula (3.72), then:

$$\lg\left(\frac{j}{T^2}\right) = \lg A + \frac{0.434}{T}\left(0.439\sqrt{E} - \frac{\phi}{k}\right) \tag{3.75}$$

When adopting zero-field simulation, the above formula can be degenerated into:

$$\lg\left(\frac{j}{T^2}\right) = \lg A - 5040\frac{\phi}{T} \tag{3.76}$$

In order to understand the above formula, a long wedge cathode is used for EM-PIC simulation, and its structure is shown in Figure 3.9. The shell is a cuboid cavity with a height and a width of 0.31 mm, a length of 1.21 mm, a wall thickness of the cavity of 0.01 mm, a length of the wedge-shaped cathode of 0.95 mm, a curvature radius of the cathode tip of about 0.2 μm and a work function of the cathode material of 2.5, 3.5 and 4.5 eV in turn.

As shown in Figure 3.10, three straight lines with different work functions can be obtained by calculating j at different temperatures and using $\lg(j/T^2)$ as the ordinate and $10,000/T$ as the abscissa. Based on its slope, the work functions calculated by formula (3.76) are 4.504, 3.511 and 2.521 eV, respectively, which are basically consistent with the original 4.5, 3.5 and 2.5 eV. In order to observe the intersection of the three straight lines

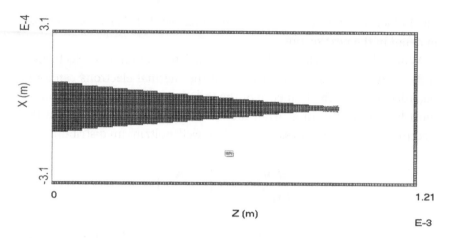

FIGURE 3.9   Device structure.

and the ordinate, the points with T of 5000 K and 10,000 K are specially selected here, but these points do not exist in the actual project because the cathode has already melted. It can be seen that the intersections of the three lines and the ordinates are all around 6.0, which is basically consistent with lg A (about 6.08).

3.7.2.2 Boundary Conditions of Field Electron Emission

As mentioned earlier, due to the Schottky effect, the applied velocity field reduces the potential barrier on the metal surface, thus reducing the effective work function and increasing the emission current. However, this result is only applicable when the external acceleration field is not strong. When the external acceleration field is strong, equation (3.72) is not applicable at all, and the emitted current is much larger than that calculated by the formula.

The tunnel effect explains this phenomenon very well. From the point of view of quantum mechanics, electrons have both volatility and granularity. For surface barriers with a certain width, electrons with energy less than the height of the barrier also have a certain probability of penetrating the barrier and escaping.

Only in the case of low field or no field and the surface barrier is very wide, the number of electrons escaping due to the tunnelling effect can be almost ignored. In the case of strong field, the height of the surface barrier decreases and so does its width. When the width of the surface barrier becomes small enough to be comparable with the length of the electron motion wave, the electrons escaping due to the tunnelling effect begin to dominate. This kind of electron emission through the barrier due to the tunnelling effect is called field electron emission.

Even at the temperature of the absolute zero, the electron penetration probability near the Fermi level is as small as 10^{-8}. However, the field emission current density is comparable to thermal electron emission of tungsten at 2800 k. The discovery of this kind of cold cathode with high current intensity solves the problem that the current density of thermal emission is not too large, which has been researched by the academic community at a large scale.

The quantitative equation of metal field emission is first deduced by Fowler and Nordheim. They assume that, first, considering a simple band of electrons, their distribution conforms to Fermi-Dirac statistics; second, considering a smooth planar metal surface, the irregularity of the atomic scale is ignored; third, considering the classical mirror force; fourth, considering the uniform distribution of the escaping work. Under this assumption, the relationship among the emission current density j(A m^{-2}), the electric field intensity E(V m^{-1}) on the surface of the emitter and the work function ϕ (eV) of the emitter material are derived

$$ j = \frac{AE_s^2}{\phi \cdot t^2(y)} \exp\left[-\frac{B \cdot v(y) \cdot \phi^{3/2}}{E_s} \right] \tag{3.77} $$

where $A = 1.54 \times 10^{-6}$ A·eV·V^2, $B = 6.83 \times 10^9$ eV$^{-3/2}$·V·m^{-1}, $t^2(y) = 1.1$, $v(y) = 0.95 - y^2$, $y = 3.79 \times 10^{-4} E_s^{1/2}/\phi$ and $E_s = \left(E_c A_c - q/\varepsilon_0 \right)/A_s$, E_s is the surface electric field, E_c is the semi-grid electric field, correspondingly, A_c is the semi-grid area, A_s is the surface area, q is the semi-grid electric quantity, and ε_0 is the dielectric constant in vacuum, which is the famous Fowler-Nordheim field emission formula.

From the Fermi energy level formula (3.71), it is not difficult to see that the electron energy of field emission is generally very small (slightly less than Fermi energy level) at the temperature of the absolute zero, so in a strong field, it can be simplified that the initial value is zero, or it can also be considered that there is a similar Maxwell distribution with the bandwidth of about 1 eV near the Fermi energy level. According to the previous analysis, at the temperature of the absolute zero, there should be no emitted electrons with energy higher than the Fermi level. There is actually no emitting cathode at the temperature of the absolute zero, so both of the above approximations have certain practical meaning.

Similarly, transforming equation (3.77) to take the logarithm is:

$$\lg\left(\frac{j}{E^2}\right) = \lg\left(\frac{A}{\phi * t^2(y)}\right) - 0.434\frac{B * v(y) * \phi^{3/2}}{E} \qquad (3.78)$$

Because the electric field E is not the same everywhere on the cathode surface, the overall current density cannot be determined by the field strength of a point. Since the field strength at each point in the tip effect has a proportional relationship with the cathode voltage, it may be assumed $E = \beta U$, so that a straight line should theoretically be obtained in a coordinate with $\lg(j)$ as the ordinate and $1/U$ as the abscissa.

Here, the wedge-shaped cathode device shown in Figure 3.9 is still used, its cathode material is tungsten (work function is 4.5 eV). It can be seen from Figures 3.11 and 3.12 that it basically conforms to the straight line when the field strength is not too large (section AB), but there will be deviation at the high field strength (section BC), which is mainly because the current density increases when the field strength is high, and the space charge effect begins to be obvious. However, the space charge effect will be more obvious when the field strength is further increased (CD segment), and the volt-ampere characteristic of emission current will show

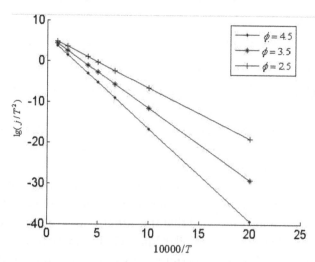

FIGURE 3.10 Thermal electron emission image of different work functions.

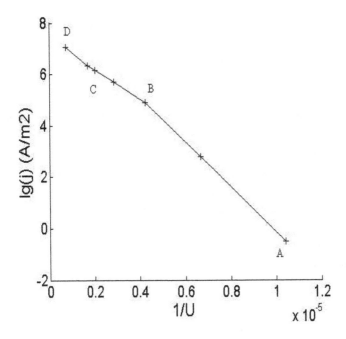

FIGURE 3.11   Field emission current density and voltage characteristic diagram.

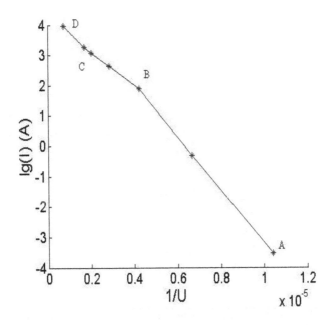

FIGURE 3.12   Field emission current and voltage characteristic diagram.

the Three-Halves Power law characteristic. In addition to the fact that the field emission is basically correct, it can also be seen that the EM-PIC method considers the effect of space charge self-consistently.

3.7.2.3 Boundary Conditions of Thermal Field Emission

In general, an emission under the conditions with the weak field strength and high temperature is described by Richardson equation, while it is described by Fowler-Nordheim equation in the case of low temperature and strong field, and there is another case, that is, the temperature and field strength are between the former two. The both kinds of emission cannot be ignored, which is generally called thermal field emission [57–59].

From the point of view of quantum mechanics, it is assumed that the transmitted electrons are all electrons near the Fermi level E_F, and the integral formula of the emission current is

$$j(T) = \frac{4\pi emkT}{h^3} \int e^{-c + \frac{E - E_F}{d}} \ln\left(1 + e^{-\frac{E - E_F}{kT}}\right) dE \qquad (3.79)$$

Simplified to be in the form

$$j(T) = j(0) \frac{\frac{\pi kT}{d}}{\sin \frac{\pi kT}{d}} \qquad (3.80)$$

where $j(0)$ is the current density at the temperature of the absolute zero degrees, k is the Boltzmann constant and T is the cathode temperature,

$$d = \frac{heE}{4\pi\sqrt{2m}|\phi|t(y_0)} \qquad (3.81)$$

$$c = \frac{8\pi\sqrt{2m}|E_F|^{3/2}\theta(y)}{3heE} \qquad (3.82)$$

where E is the field strength of the cathode, ϕ is the work function of the metal and $t(y_0)$ is approximately 1.

The condition for formula (3.80) to be established is $d > kT$, simplified to $E > 8.83 * 10^5 \phi^{1/2} T$, which is the limit of field emission. When the temperature is higher than this limit, it means that more high-level electrons escape, and E is not too large (because the melting point of the metal limits the temperature cannot increase indefinitely). In this case, the thermal electron emission should be mainly considered, which is defined by equation (3.82). If the temperature is very low, the emission generated is mainly field emission (if field emission can be generated), then it can be considered by equation (3.77).

When T approaches zero, equation (3.80) degenerates into equation (3.77), which is question on the field emission at the absolute zero degree. When T is not too large, equation (3.80) can be expanded as

$$j(T) = j(0)\left[1 + \frac{1}{6}\left(\frac{\pi kT}{d} \right)^2 + \cdots \right] \qquad (3.83)$$

$$\frac{\pi kT}{d} = 2.77 \times 10^2 \frac{T\sqrt{\phi}}{E} \qquad (3.84)$$

which can be regarded as a correction to the absolute zero current density when T becomes larger. Taking a tungsten cathode as an example here, when the surface field strength reaches 4×10^9 $V\,m^{-1}$ and at room temperature ($T = 300$ K), $\pi kT/d$ is approximately equal to 0.4, which can be seen from equation (3.83), $j(T) = 1.03 \times j(0)$. When the temperature is increased to 1000 K, $\pi kT/d$ is approximately equal to 1.5. From equation (4.51), $j(T) = 1.5 \times j(0)$. It can be seen that the thermal electron emission is only 3% of the total when the temperature rises from $T = 0$ K to $T = 300$ K, and one third of the electrons are emitted by the thermal electrons, from high-energy level escaped, when the temperature rises to 1000 K [57–59].

Generally, the electron energy emitted by thermionic emission will be several electron volts higher than the Fermi level, the total energy is still relatively small, and it can still be considered to be zero under normal circumstances. In the case that the electron initial velocity and energy distribution must be considered, the analysis of the thermal electron emission can be similar.

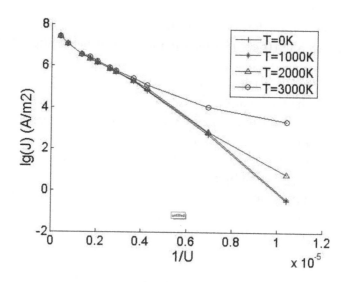

FIGURE 3.13   Effect of temperature on current density in thermal field emission.

Here, the wedge-shaped cathode device is still used, its cathode material is tungsten (work function is 4.5 eV). It can be clearly seen from Figure 3.13 that the temperature has little effect on the current density when the temperature is less than 1000 K, and the effect increases significantly when the temperature becomes higher. In the strong electric field, the effect of temperature is also weak, this is mainly because there are too many electrons that can escape through the barrier, and the effect of hot electron emission becomes negligible. This is in good agreement with the theory [57–59].

For thermal electron emission and thermal field emission, the emission model is based on the quantum mechanics theory. Because the physical mechanism is very complex, it mainly studies the emission of metals, which has good applicability to common metals such as molybdenum and tungsten. The field emission of the cathode, such as semiconductors, diamond, zinc oxide, and carbon nanotube, is all developed on this basis, and some of them can be directly simulated using the above model. Here, due to the limitation of space, we will not discuss them one by one. For those who are interested, please refer to the literature [57–59].

3.7.2.4 Space Charge Limited Emission
In recent years, increasing attention has been paid to space charge limited current [60–64], which has been widely used in many high current pulse and high power microwave source devices. One of the most commonly

used space charge limited emission is plasma field emission, also known as explosive electron emission [65]. It is a process of ionization, heating and melting to explosion at the tip of the cathode surface caused by the increase of field emission current. This process has a large number of ions and electrons entering the space, so it also belongs to the space charge limited state.

The core idea of space charge limited emission [66] is to generate a strong current from the cathode under a strong electric field, which can make the field strength of the cathode surface zero. Because the volt-ampere characteristic of this flow in the diode exhibits the Child-Langmuir relationship (i.e., the Three-Halves Power law), it is also called Child emission. Applying this relation on the discrete grid, the current density on the first grid of the emitter surface can be obtained:

$$J = \frac{4\varepsilon_0}{9\Delta x^2} \left(\frac{2e}{m} \right)^{1/2} (-E\Delta x)^{3/2} \tag{3.85}$$

where E is the electric field of the first grid midpoint on the cathode surface, and Δx is the length from the cathode surface to the first grid midpoint. Obviously, this method is to realize the emission process by virtual a diode on the cathode surface.

Another method can also describe the emission model, that is, to apply the Gauss theorem to a volume in the simulation area. In calculation, this concept is to import particles by controlling the volume of the cathode surface grid and applying Gauss theorem to this volume, the following formula can be obtained:

$$\nabla \cdot E = \rho / \varepsilon_0 \tag{3.86}$$

$$\iiint \nabla \cdot E d^3 x = \iiint \rho / \varepsilon_0 \, d^3 x \tag{3.87}$$

$$\oiint_\Omega E d^2 x = Q / \varepsilon_0 \tag{3.88}$$

The amount of electricity that needs to be added to the volume is:

$$Q_{add} = \varepsilon_0 \oiint_\Omega E d^2 x - Q_{old} \tag{3.89}$$

where Q_{old} is the amount of charge already in the volume.

Generally, this formula is applied to half of the grid on the emitting surface, and a large number of charged macroparticles are injected into the simulation area, thereby satisfying that the electric field on the emitting surface is zero, and then giving these macroparticles a certain initial speed is sufficient.

In addition, due to the practicality and importance of space charge limiting current, many researchers have made some improvements and amendments to their theories and models. Those interested can refer to the literature [67–69]. In addition to the temperature- or space-limited emission flow, there may also be a composite limited flow, that is, temperature space charge limited emission. This is because the cathode under normal conditions is difficult to keep at low temperature, let alone at the temperature of the absolute zero. Therefore, when the voltage between anode and cathode is not too high, the emission of hot cathode often lies between the temperature-limiting flow and the space limiting flow. This process can be expressed as follows:

$$J_c = \frac{J_{SC} + J_{TL}}{J_{SC} J_{TL}} \tag{3-90}$$

where J_{SC} and J_{TL} represent space charge limited flow and temperature charge limited flow, respectively, which can be described by the expression of space charge limited flow formula (3.85) and the expression of thermal electron emission formula (3.70) under acceleration field, respectively.

The initial velocity of electron emission from the cathode surface can be described as follows:

$$v_i = c \sqrt{\frac{\phi(\phi + 2\lambda)}{(\phi + \lambda)^2}} \tag{3.91}$$

$$\theta_i = \cos^{-1}\left(\frac{E_z}{E}\right) \tag{3.92}$$

where $\lambda = mc^2/e$, c is the speed of light, θ_i is the angle between the initial speed and the normal direction, E_z is the normal electric field and E is the cathode surface electric field. This has a certain guiding significance in the general consideration of the restricted current generated by the hot cathode.

Under the low voltage between cathode and anode, the electron emission of hot cathode is more limited by space charge effect. When the voltage between cathode and anode is high, the temperature-limiting flow begins to occupy the main position. In this case, it can only be expressed by equation (3.72). When the voltage between the anode and cathode rises again and breaks through the field emission limit, the field emission formula or the thermal field emission formula should be used to describe the process. When the voltage rises again, the space charge limited effect will dominate. It can be seen that space charge effect and temperature-limiting effect exist in any case, but in some cases, these effects are not obvious enough. However, when the voltage drops to the opposite direction, i.e. in the rejection field, the thermionic cathode still has electron emission. Because this kind of situation is rarely used, we will not discuss it in detail here.

3.8 FIELD BOUNDARY CONDITION

The microwave components of spacecraft are usually characterized with small electrical size and high microwave frequency. In general, due to the large amount of calculation, the calculation area cannot be very large, which requires the boundary condition of truncating the calculation area. In addition to the most common boundaries, such as conductor, excitation, and absorption, symmetric boundary condition is also one of the important boundary conditions for simplified calculations. These types of boundary conditions are discussed separately below.

3.8.1 Conventional Field Boundary Condition

The conventional boundary conditions of the electromagnetic field include conductor boundary conditions, medium boundary conditions, and symmetric boundary conditions. The processing of these types of boundary conditions is relatively simple. It is only necessary to use the corresponding formula to obtain the new field value based on the current electromagnetic field value. Each will be briefly described below.

3.8.1.1 Conductor and Dielectric Boundary Condition

Conductor boundary conditions can be divided into perfect conductor boundary condition and non-perfect conductor boundary condition.

The processing of perfect conductor boundary conditions is very simple, which is only followed that the electric field inside the conductor and the tangential electric field on the surface of the conductor are zero. In term of the numerical realization, some electric field components of the conductor are corrected to zero again after each step of electromagnetic field calculation.

For the boundary condition of non-perfect conductor or medium, the standard FDTD formula in Section 3.4.3 can be used. For the calculation of the simple format in Section 3.4.3, simple correction is needed. According to Ohm's law, a term σE will be added to the right of equation (3.3), which results in the following form:

$$\nabla \times H = \varepsilon \frac{\partial E}{\partial t} + \sigma E + J \tag{3.93}$$

By further transformation, then

$$\frac{E^{n+1} - E^n}{\Delta t} + \frac{\sigma E^{n+1}}{\varepsilon} = \frac{1}{\varepsilon}(\nabla \times H - J) \tag{3.94}$$

where E^n is the electric field value at the time of n, E^{n+1} is the electric field value at the time of $n + 1$. Simplified to be in the form

$$\left(1 + \frac{\sigma \Delta t}{\varepsilon}\right) E^{n+1} = E^n + \frac{\Delta t}{\varepsilon}(\nabla \times H - J) \tag{3.95}$$

After simplifying, we have

$$E_{temp}^{n+1} = \left(1 + \frac{\sigma \Delta t}{\varepsilon}\right) E^{n+1} = E^n + \frac{\Delta t}{\varepsilon}(\nabla \times H - J) \tag{3.96}$$

After extracting the front and back items, we have

$$E_{temp}^{n+1} = E^n + \frac{\Delta t}{\varepsilon}(\nabla \times H - J) \tag{3.97}$$

Obviously, this equation is completely consistent with the simplified central difference scheme in Section 3.4.3. So E_{temp}^{n+1} can first be calculated by using this difference scheme, which is amended according to the finite conductivity of each grid in the calculation area, that is, dividing all electric field components by a coefficient $(1 + \sigma \Delta t / \varepsilon)$, and then the new electric field value E^{n+1} can be obtained.

The specific steps of numerical simulation are that an array is established for the relative dielectric coefficient, relative permeability and conductivity of each grid point, then the initial value on all grid points is given, where the default value of relative dielectric coefficient and relative permeability is 1, and the initial value of conductivity is 0. The relative dielectric coefficient, relative permeability and electrical conductivity of the dielectric or non-perfect conductor are read, with which the initial value of the area is corrected. The modification of the dielectric constant or conductivity on the grid added in the previous step will result in a sudden change on the boundary of these values on the grid, which may cause the inestimable calculation results. Therefore, the values of these three arrays are, respectively, weighted on the grid to make it smooth. The electromagnetic field value in the region can be obtained directly through the iteration of the standard difference formula, and then all the electric field components are divided by a coefficient $(1 + \sigma \Delta t / \varepsilon)$ to be modified by equation (3.96) because the initial value of conductivity is set as 0 in the vacuum region, and the coefficient above is degenerated to 1, and equation (3.93) is naturally degenerated to the difference format in vacuum. Obviously, this method is valid for conventional dielectric or non-perfect conductor.

This kind of simplification and modification is very useful in many programs, especially in some complex large-scale programs (such as the micro discharge electromagnetic particle simulation that will be involved in the future), which can play a role of simple program structure and reducing the amount of calculation. For FDTD algorithm, we should consider the flexible use of the algorithm without affecting the accuracy of calculation and not stick to a specific format or method.

3.8.1.2 Symmetry Boundary Condition

Symmetrical boundary conditions can be subdivided into axisymmetric boundary condition, mirror symmetric boundary condition and periodic symmetric boundary condition. The schematic diagram of these three symmetrical boundary conditions is shown in Figure 3.14. In the case of axisymmetry and mirror symmetry, according to the type of

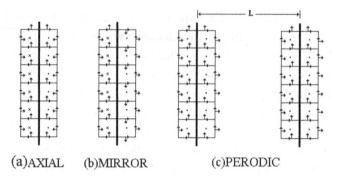

FIGURE 3.14   (a–c) The schematic diagram of these three symmetrical boundary conditions.

electromagnetic field components on one side of the symmetrical boundary and symmetrical boundary conditions, the electromagnetic field components on the other side can be obtained by directly copying some field components or inversing some other field components; in the case of periodic symmetry, the field value of the corresponding position after one period separation is directly copied from the field value of the starting position.

Axisymmetric boundary condition is generally used in cylindrical coordinate system, and only in the position where the radial coordinate value is zero. In 2D and 3D EM-PIC simulation, axisymmetric boundary condition is needed as long as the simulation area contains axes. An example of 2D axisymmetric boundary condition is given in the figure.

In the axisymmetric boundary, usually only the field in a section containing the symmetry axis is calculated to reduce the amount of calculation, so the E_z on the axis needs to be treated separately. According to Ampere circuital theorem, the FDTD difference scheme of axis field can be obtained:

$$E_z^{n+1}\left(0, j+\frac{1}{2}\right)= E_z^n\left(0, j+\frac{1}{2}\right)+\frac{4\Delta t}{\varepsilon \Delta r} H_\varphi^{n+1/2}\left(\frac{1}{2}, j+\frac{1}{2}\right) \quad (3.98)$$

The mirror symmetry boundary condition is generally used to the case with the known symmetry of the simulated structure on both sides of the boundary. For example, due to the symmetry of some modes excited in the waveguide, it is unnecessary to simulate the whole region. The mirror

FIGURE 3.15   Schematic diagram of periodic boundary condition.

symmetry boundary can be used to partition from the middle, and only half or smaller regions should be simulated to obtain the electromagnetic field distribution of the whole region.

Periodic symmetric boundary condition is generally used for structures with periodic characteristics. For example, when analysing a long periodic slow wave structure, only one segment can be selected for analysis, and the start and end of this segment can be taken as two symmetrical surfaces of the periodic symmetric boundary condition. Then, the field values at both ends will be corrected again in each iteration to make them equal to each other, so as to achieve the purpose of simulating an approximately infinite long structure. This example is shown in Figure 3.15.

Another important use of periodic symmetric boundary condition is to deal with circular structures. As shown in Figure 3.16, in the 2D polar coordinate system, since the position coincide of 0° and 360° of the structure in the angular direction, the electromagnetic field values in the two positions should be completely equal in physics, but the two positions are calculated separately in the numerical value, so it is necessary to set the periodic symmetric boundary condition in the two positions, and each iteration is modified once.

3.8.2 Excitation Source Boundary Condition

The excitation source boundary condition is the most important electromagnetic field boundary condition in the EM-PIC method. This type of boundary condition is used to provide the source of the electromagnetic field, which is the basis for studying all electromagnetic problems. This section introduces the time domain and frequency domain characteristics of several commonly used time harmonic and pulse sources in FDTD.

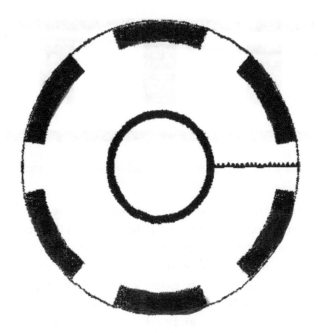

FIGURE 3.16   Periodic symmetric boundary of circular structure.

3.8.2.1 Time Harmonic Field Source

In order to calculate the electromagnetic problems in the case of time harmonic field by the FDTD method, it can be assumed that the incident field is

$$E_i(t) = \begin{cases} 0 & t < 0 \\ E_0 \sin(\omega t) & t \geq 0 \end{cases}, \tag{3.99}$$

which is a sine function starting at 0 obviously. The field value suddenly generated at time 0 not only easily provokes additional electromagnetic components in the calculation but also differs from the actual gradual field construction process, so generally the field establishment process needs to be considered. The rise of the electromagnetic field from zero to the steady state of the above formula often takes several cycles, which is related to the Q value of the cavity. In order to shorten the steady-state settling time and reduce the impact effect, a switching function is often introduced. Then, the above formula can be expressed as

$$E_i(t) = E_0 U(t)\sin(\omega t) \tag{3.100}$$

TABLE 3.3   Common Switching Function Table (t_0 is a Constant Representing Time)

Switching Function	Functional Form
Step function	$U(t)=\begin{cases} 0 & t<0 \\ 1 & t\geq 0 \end{cases}$
Inverse exponential function	$U(t)=\begin{cases} 0 & t<0 \\ 1-\exp\left(-7t/t_0\right) & 0\leq t<t_0 \\ 1 & t\geq t_0 \end{cases}$
Ramp function	$U(t)=\begin{cases} 0 & t<0 \\ \left(t_0-t\right)/t_0 & 0\leq t<t_0 \\ 1 & t\geq t_0 \end{cases}$
Raise cosine function	$U(t)=\begin{cases} 0 & t<0 \\ 0.5\left[1-\cos\left(\pi t/t_0\right)\right] & 0\leq t<t_0 \\ 1 & t\geq t_0 \end{cases}$

where $U(t)$ is the switching function. The following table gives several common switch functions for readers to refer to.

3.8.2.2 Pulse Source

The spectrum of the pulse source usually has a certain bandwidth. Understanding the pulse source and its spectrum characteristics is very important for FDTD calculation. The following focuses on several commonly used pulse sources.

All the above pulses, except Gauss pulse, are zero at the truncation, which are finite base functions. They have good smoothness, and there is no DC component in differential Gaussian pulse and modulated Gaussian pulse.

3.8.2.3 Waveguide Excitation Source

When the waveguide excitation source is simulated by FDTD, it is necessary to ensure that the characteristics of the source are consistent with the actual physical model as much as possible, which requires that the scattering field at the boundary of the excitation source must be considered reasonably. At present, except for some simple applications using forced excitation source, most of the introduction methods of the excitation source are based on the total field/scattering field system. The traditional

TABLE 3.4 Common Pulse Source Expression

Pulse Source	Time Domain Expression	Frequency Domain Expression	Typical Characteristics
Gaussian pulse	$E_i(t)=\exp\left(-\dfrac{4\pi(t-t_0)^2}{\tau^2}\right)$	$E_i(f)=\dfrac{\tau}{2}\exp\left(-j2\pi f t_0-\dfrac{\pi f^2\tau^2}{4}\right)$	The peak value of pulse appears at $t=t_0$
Raise cosine pulse	$E_i(t)=\begin{cases} 0.5\left[1-\cos\left(\dfrac{2\pi t}{\tau}\right)\right] & 0\le t\le\tau \\ 0 & \text{other} \end{cases}$	$E_i(f)=\dfrac{\tau\exp(-j\pi f\tau)}{1-f^2\tau^2}\dfrac{\sin(\pi f\tau)}{\pi f\tau}$	τ is the pulse base width
Differential Gaussian pulse	$E_i(t)=\dfrac{t-t_0}{\tau}\exp\left(-\dfrac{4\pi(t-t_0)^2}{\tau^2}\right)$	$E_i(f)=\dfrac{-j\tau^2 f}{8}\exp\left(-j2\pi f t_0-\dfrac{\pi f^2\tau^2}{4}\right)$	Zero-frequency component not included
Modulated Gaussian pulse	$E_i(t)=-\cos(\omega t)\exp\left(-\dfrac{4\pi(t-t_0)^2}{\tau^2}\right)$	$E_i(f)=\dfrac{\tau}{4}\exp\left[-\dfrac{\pi(f-f_0)^2\tau^2}{4}\right]\exp\left[-j2\pi(f-f_0)t_0\right]$ $+\dfrac{\tau}{4}\exp\left[-\dfrac{\pi(f+f_0)^2\tau^2}{4}\right]\exp\left[-j2\pi(f+f_0)t_0\right]$	The centre frequency is $f_0=\omega/2\pi$

method of setting up waveguide excitation source is that the forced excitation source can be easily established by directly assigning the required time variation form to the specific electric field or magnetic field component. The forced excitation source is simple in form and easy to use. It is used to deal with many engineering problems without the consideration of the scattering field. This is easy to cause false reflection at the boundary of the excitation source, resulting in data error. Therefore, the corresponding improvement should be carried out on the basis of the forced excitation source, so that the scattered field can smoothly leave the simulation area through the boundary of the excitation source, without being reflected and affecting the field data in the simulation area.

Generally, two methods are used to overcome the false reflection caused by the forced excitation source: one is to remove the excitation source before the scattered wave from the structure reaches the grid point of the excitation source when the excitation pulse almost decays to zero, which is only applicable to the pulse excitation source with a narrow pulse width; the other is to set a current excitation source to excite an electric field, which is equivalent to the required electric field excitation form. This method can simulate the sine wave excitation source, Gaussian pulse source or sampling function excitation source and so on, but it is difficult to simulate the more complex waveguide excitation source, and the equivalent current source cannot be placed at the boundary of the simulation area.

Taking the electric field excitation as an example, the subscripts "t", """" and "s" are used to represent "total field", "incident field" and "scattering field", respectively. The relationship among the three is as follows:

$$E_t = E_i + E_s \qquad (3.101)$$

The calculation model is as follows: assuming that the incident port is located in the plane S, the incident field propagates in the positive direction of the X axis, the scattering field propagates in the negative direction of the X axis, plane E is parallel to plane S, the vertical distance is a grid step, and M is the parallel plane between E and S. Then, the weight expression of scattering field at time t in plane M can be obtained from the above formula

$$E_s(M,t) = \alpha E_s(M,t) + \beta E_s(E,t) \qquad (3.102)$$

where the weight factor $\alpha = 1 - v_p \Delta t / \Delta x$ and $\beta = 1 - \alpha = v_p \Delta t / \Delta x$ represents the proportion of scattering fields on plane S and E, respectively. It can be seen that the weight factor is inversely proportional to the distance from this plane to plane M. In this case, the scattering field at time $t + \Delta t$ on the plane S is equal to the scattering field at time t on the plane M, i.e.

$$E_s(S, t + \Delta t) = E_s(M, t) \tag{3.103}$$

The latter item in formula (3.66) is not difficult to get from

$$E_s(E, t) = E_t(E, t) - E_i(E, t) \tag{3.104}$$

In the above formula, the first term on the right can be obtained directly, and the second term can be calculated from the propagation characteristics of the wave: let the time for the incident field to propagate from plane S to plane E be $\Delta t'$, when $t \geq \Delta t'$, the incident field on plane E at time t is equal to the incident field on plane S at that time $t - \Delta t'$; when $t < \Delta t'$, the incident field has not yet propagated to plane E, so the value is zero, and hence

$$E_i(E, t) = \begin{cases} 0 & 0 \leq t < \Delta t' \\ E_i(S, t - \Delta t') & t \geq \Delta t' \end{cases} \tag{3.105}$$

where $\Delta t' = \Delta x / v_p$, that is, the grid step in the propagation direction divided by the phase velocity.

Combining the above formulas, iterative formulas for the total field and scattered field at the incident port can be obtained

$$E_s(S, 0) = 0, \tag{3.106}$$

$$E_t(S, t) = E_i(S, t) + E_s(S, t) \tag{3.107}$$

$$E_s(S, t + \Delta t') = \begin{cases} \alpha E_s(S, t) + \beta E_t(E, t) & 0 \leq t < \Delta t' \\ \alpha E_s(S, t) + \beta E_t(E, t) - \beta E_i(S, t - \Delta t') & t \geq \Delta t' \end{cases} \tag{3.108}$$

The specific implementation process is: first, the field value in the whole simulation area is initialized to zero, according to the iterative formula, the total field value on the boundary is updated by the incident field function formula and the scattering field value on the boundary at the current time, and then the scattering field value on the boundary of the excitation source at the next time is calculated according to the iterative formula for the next calculation of the total field on the boundary; the calculated total field value on the boundary is introduced into the whole FDTD iteration to update the field value in the whole simulation area through calculation, and repeat the iteration until the end of the simulation. It is worth mentioning that the derivation here considers the case where the excitation source is placed in the flat waveguide in the vacuum environment, which can meet the needs of the vast majority of cases, so this chapter will not be further discussed. If some readers want to further study the algorithm implementation in more complex situations, they can refer to the relevant literature [70,71].

3.8.3 Electromagnetic Wave Absorption Boundary Condition

In FDTD, the electromagnetic wave absorption boundary condition is a very important basic element. Because the boundaries of many practical problems (such as radiation and scattering of electromagnetic fields) are open, the electromagnetic field will occupy infinite space. However, the computer memory is always limited, so it can only simulate a limited space. In actual simulation, it is necessary to simulate the problem of infinite space in a limited space. Therefore, setting the boundary of the simulation area becomes an important task. The key of this problem is how to set the boundary conditions so that the electromagnetic wave can be absorbed by the boundary without reflection.

Among the developed electromagnetic wave absorbing boundary conditions, the most accurate one is the PML absorbing boundary condition, which was first proposed by Berenger [72,73] in 1994. Among several PML absorbing boundary conditions, Gedney [74] perfectly matched layer (GPML) absorbing boundary condition is easy to understand and realize, and most widely used, whose main idea is to set an uniaxial high loss isotropic medium with a certain thickness at the boundary, so as to achieve the non-reflection attenuation and absorption of electromagnetic wave.

First, since there are no sources inside the absorption boundary, the current density vector in formula (3.3) of Maxwell equations can be ignored. According to formula (3.2) and formula (3.3), we can get:

$$\nabla \times \boldsymbol{E} = -\frac{\partial \boldsymbol{B}}{\partial t} = -j\omega \boldsymbol{B} = -j\omega\mu_0\mu_r\ddot{\mu}\boldsymbol{H} \qquad (3.109)$$

$$\nabla \times \boldsymbol{H} = -\frac{\partial \boldsymbol{D}}{\partial t} = -j\omega \boldsymbol{D} = -j\omega\varepsilon_0\varepsilon_r\ddot{\varepsilon}\boldsymbol{D} \qquad (3.110)$$

where ε_r and μ_r are relative permittivity and relative permeability, respectively, tensor $\ddot{\varepsilon}$ and $\ddot{\mu}$ are defined as

$$\ddot{\varepsilon} = \begin{bmatrix} a & 0 & 0 \\ 0 & a & 0 \\ 0 & 0 & b \end{bmatrix} \qquad (3.111)$$

$$\ddot{\mu} = \begin{bmatrix} c & 0 & 0 \\ 0 & c & 0 \\ 0 & 0 & d \end{bmatrix} \qquad (3.112)$$

The wave equation can be derived from the above coupled wave equation, and the non-reflection condition can be obtained:

$$a = c = \frac{1}{b} = \frac{1}{d} \qquad (3.113)$$

The incident wave will transmit in the matched layer without reflection as long as this condition is satisfied. This condition does not change with the frequency, phase, polarization and incident angle of the incident wave. Because of this characteristic, lossy perfect matching layer is often introduced to make electromagnetic wave enter into it without loss, and quickly collapse, so as to achieve the purpose of simulating open boundary. In order to make the matching layer have large loss, let $a = 1 + \sigma_z/j\omega\varepsilon_0$, and tensor $\ddot{\varepsilon}$ and $\ddot{\mu}$ can be expressed as

$$\ddot{\varepsilon} = \ddot{\mu} = \begin{bmatrix} 1+\sigma_z/j\omega\varepsilon_0 & 0 & 0 \\ 0 & 1+\sigma_z/j\omega\varepsilon_0 & 0 \\ 0 & 0 & \dfrac{1}{1+\sigma_z/j\omega\varepsilon_0} \end{bmatrix} \quad (3.114)$$

Obviously, when σ_z approaches zero, the matching layer medium degenerates into isotropic medium. By substituting the above formula into formula (4.29) and (4.30), and taking the rectangular coordinate system as an example, the difference formula of electromagnetic field calculation in GPML absorbing boundary area can be obtained:

$$\begin{bmatrix} \dfrac{\partial H_z}{\partial y} - \dfrac{\partial H_y}{\partial z} \\ \dfrac{\partial H_x}{\partial z} - \dfrac{\partial H_z}{\partial x} \\ \dfrac{\partial H_y}{\partial x} - \dfrac{\partial H_x}{\partial y} \end{bmatrix} = j\omega\varepsilon_0\varepsilon_r \begin{bmatrix} 1+\sigma_z/j\omega\varepsilon_0 & 0 & 0 \\ 0 & 1+\sigma_z/j\omega\varepsilon_0 & 0 \\ 0 & 0 & \dfrac{1}{1+\sigma_z/j\omega\varepsilon_0} \end{bmatrix} \begin{bmatrix} E_z \\ E_y \\ E_x \end{bmatrix}$$

$$(3.115)$$

$$\begin{bmatrix} \dfrac{\partial E_z}{\partial y} - \dfrac{\partial E_y}{\partial z} \\ \dfrac{\partial E_x}{\partial z} - \dfrac{\partial E_z}{\partial x} \\ \dfrac{\partial E_y}{\partial x} - \dfrac{\partial E_x}{\partial y} \end{bmatrix} = j\omega\mu_0\mu_r \begin{bmatrix} 1+\sigma_z/j\omega\varepsilon_0 & 0 & 0 \\ 0 & 1+\sigma_z/j\omega\varepsilon_0 & 0 \\ 0 & 0 & \dfrac{1}{1+\sigma_z/j\omega\varepsilon_0} \end{bmatrix} \begin{bmatrix} H_z \\ H_y \\ H_x \end{bmatrix}$$

$$(3.116)$$

Obviously, the first two expressions in the above two formulas are completely consistent with the FDTD expressions in the conventional isotropic media, and can be calculated by the standard difference scheme. The third expression needs to be treated separately. Taking the electric field treatment as an example, it may be assumed that

$$\bar{E}_z = \frac{1}{1+\sigma_z/j\omega\varepsilon_0} E_z \quad (3.117)$$

Conversion to time domain expression, and hence

$$\frac{\partial \overline{E}_z}{\partial t} + \frac{\sigma_z}{\varepsilon_0} \overline{E}_z = \frac{\partial E_z}{\partial t} \qquad (3.118)$$

Then, its central difference format is

$$E_z^{n+1} = E_z^n + \left(1 + \frac{\sigma_z \Delta t}{2\varepsilon_0}\right) \overline{E}_z^{n+1} - \left(1 - \frac{\sigma_z \Delta t}{2\varepsilon_0}\right) \overline{E}_z^n \qquad (3.119)$$

In order to avoid reflection on the surface of the matching layer, it is required that the value of σ_z gradually increases from the surface. Generally, it is defined as

$$\sigma_z(z) = \sigma_{max} \frac{|z - z_0|^m}{d_z^m} \qquad (3.120)$$

where d_z is the thickness of matching layer and z_0 is the surface coordinate of matching layer. In general experience, $m = 4$, $dz = (8-10)\Delta z$ and $\sigma_{max} = \dfrac{m+1}{150\pi\Delta\sqrt{\varepsilon_r}}$ are often taken.

3.9 STABILITY CONDITION

In FDTD, the discrete method is used to solve continuous problems, which will inevitably introduce errors. Therefore, whether the calculation area or the matching layer area has its stability conditions in the FDTD calculation:

1. The calculation time step cannot be too large. The time step and space step must meet the following stability conditions, which are the well-known Courant stability conditions:

$$c\Delta t \leq \frac{1}{\sqrt{\dfrac{1}{(\Delta x)^2} + \dfrac{1}{(\Delta y)^2} + \dfrac{1}{(\Delta z)^2}}} \qquad (3.121)$$

where c is the speed of light in vacuum. If there is a non-uniform medium or a non-uniform grid in the simulation area, then the extreme values of the speed of light and the size of the grid must be considered.

It can be seen from this condition that the time interval must be less than or equal to the time of light passing through the Yee cell in vacuum. Because the velocity of particles cannot exceed the speed of light, the motion distance of particles in time interval can't be larger than a mesh. This condition is also a dependent condition for the design of field particle interaction algorithm.

2. Cell size cannot be too large. In addition to affecting the physical structure of small size, it will also increase the discreteness of large field, the jumping of data and the numerical noise. Too large grid will greatly improve the numerical noise and may get bad results in the case of more high-frequency modes. This is the requirement of numerical dispersion for spatial dispersion. Generally, the maximum mesh should be less than 1/12 of the minimum wavelength. This limitation is generally more severe than Courant stability conditions. Similarly, the discrete time is less than 1/12 of the period of plane electromagnetic wave. The special attention should be paid to these conditions in some media calculation.

3. Anisotropic stability condition of electromagnetic wave. It requires that the anisotropy caused by the difference approximation can be ignored only when the spatial dispersion (i.e., cell size) is less than one eighth of the wave length of electromagnetic wave, that is, a wavelength must include more than eight meshes. This is also an important point in the FDTD method, especially in the simulation of microwave and terahertz band. It is generally combined with the stability conditions of Clause 2 and requires that a wavelength must contain more than 12 meshes.

4. The variation of adjacent meshes shall not exceed 25%, and the aspect ratio of meshes shall not exceed 5. In general, the variation of adjacent meshes can be correctly obtained by the function partition algorithm, while the initial mesh aspect ratio needs to be given by the user himself. In the simulation, attention should be paid to this stability condition, which is a general condition obtained from the numerical stability.

The stability, dispersion, anisotropy and mesh in different directions calculated by the plane wave difference method are studied.

These characteristics are not determined by the calculation area or electromagnetic field characteristics, but by the approximate calculation of the difference method. In general, in order to ensure the stability and credibility of the calculation, a certain margin should be left in the application of the above stability conditions, even if it is just in the stability condition.

3.10 SUMMARY

Based on the FDTD method and PIC method, this chapter introduces the EM-PIC method in detail. For the multipactor phenomenon in spacecraft, whether waveguide, divider or other microwave components can be simulated by the method of this chapter. In addition, in this chapter, the possible particle emission in microwave environment, from low temperature to high temperature, from low field strength to high field strength, and the boundary conditions is relatively sufficiently studied in detail. In the case that some unequal charges are serious or the temperature rises obviously, and the initial particles need to be considered, a specific setting is required.

REFERENCES

1. Buneman O. Dissipation of currents in ionized media. *Physical Review,* 1959, 115:507.
2. Dawson JM. One-dimensional plasma model. *Physics of Fluids,* 1962, 5:445.
3. Birdsall CK, Langdon AB. *Plasma Physics via Computer Simulation.* New York: McGraw-Hill, 1985.
4. Hockney RW, Eastwood JW. *Computer Simulation Using Particles.* New York: McGraw-Hill, 1991.
5. Vahedi V, DiPeso G. Simultaneous potential and circuit solution for two-dimensional bounded plasma simulation codes. *Journal of Computational Physics,* 1997, 131(1):149–163.
6. Verboncoeur JP, Langdon AB, Gladd NT. An object-oriented electromagnetic PIC code. *Computer Physics Communication,* 1995, 87(1):199–211.
7. Blahovec JD, Bowers LA, Luginsland JW, et al. 3-D ICEPIC Simulations of the relativistic klystron oscillator. *IEEE Transactions on Plasma Science,* 2000, 28(3):821–829.
8. Liewer PC, Decyk VK. A general concurrent algorithm for plasma particle-in-cell codes. *Journal of Computational Physics,* 1989, 85(2):302–322.
9. Martino B, Briguglio S, Vlad G, et al. Parallel PIC plasma simulation through particle decomposition techniques. *Parallel Computing.* 2001, 27(3):295–314.

10. Verboncoeur JP. Symmetric spline weighting for charge and current density in particle simulation. *Journal of Computational Physics*, 2001, 174(1):421–427.

11. Surendra M, Graves DB, Morey IJ. Electron heating in low-pressure RF glow discharges. *Applied Physics Letters*, 1990, 56(11):1022–1025.

12. McDaniel EW, Mitchell JBA, Rudd ME. *Atomic Collisions: Heavy-Particle Projectiles*. New York: John Wiley & Sons, Inc., 1993.

13. Christlieb AJ, Krasny R, Verboncoeur JP. A grid-free treecode field solver for plasma simulations with application to a confined electron column in a Penning-Malmberg trap. *Computer Physics Communication*, 2004, 164(1):306–310.

14. Takizuka T, Abe H. A binary collision model for plasma simulation with a particle code. *Journal of Computational Physics*, 1977, 25(3):205–219.

15. Nanbu K. Momentum relaxation of a charged particle by small-angle coulomb collisions. *Physical Review E*, 1997, 56(6):7314.

16. Gopinath VP, Verboncoeur JP, Birdsall CK. Multipactor electron discharge physics using an improved secondary emission model. *Physics of Plasmas*, 1998, 5(5):1535–1540.

17. Verboncoeur JP. A digital filtering scheme for particle codes in curvilinear coordinates, 28th IEEE ICOPS, Las Vegas, N.V., 2001.

18. Lawler JE, Parker GJ, Hitchon WN. Radiation trapping simulations using the propagator function method. *Journal of Quantitative Spectroscopy & Radiative Transfer*, 1993, 49(6):627–664.

19. Lee HJ, Verboncoeur JP. A radiation transport coupled particle-in-cell simulation: Part I. Description of the model. *Physics of Plasmas*, 2001, 8(6):3077–3088.

20. Lee HJ, Kim HC, Yang SS, et al. Two-dimensional self-consistent radiation transport model for plasma display panels. *Physics of Plasmas*, 2002, 9(6):2822–2830.

21. Luginsland JW, Antonsen TA, Verboncoeur JP, et al. *Computational Techniques High-Power Microwave Sources and Technologies*. New York: IEEE Press, 2001.

22. Bowers KJ. Accelerating a particle-in-cell simulation using a hybrid counting sort. *Journal of Computational Physics*, 2003, 173(2):393–411.

23. Bowers KJ. Speed optimal implementation of a fully relativistic 3d particle push with charge conserving current accumulation on modern processors, 18th International Conference on The Numerical Simulation of Plasmas, N. Falmouth, MA, 2003.

24. Matsumoto H, Omura Y. *Computer Space Plasma Physics: Simulation Techniques and Software*. Tokyo: Terra Scientific Publishing Company, 1993.

25. Goplen B, Ludeking L, Smithe D, et al. User configurable MAGIC for electromagnetic PIC calculations. *Computer Physics Communication*, 1995, 87(1):54–86.

26. Tarakanov VP. *User's Manual for Code KARAT.* Springfield, VA: Berkeley Research Associates, Inc., 1999.

27. Kory CL. Three-dimensional simulation of helix traveling-wave tube cold-test characteristics using MAFIA. *IEEE Transactions on Electron Devices,* 1996, 45(8):1317–1319.

28. Clark RE., Hughes TP. *LSP User's Manual and Reference for LSP Version 8.7.* Newington, VA: ATK Mission Systems, Inc., 2005.

29. Mardahl P, Greenwood A, Murphy T, et al. Parallel performance characteristics of ICEPIC, User Group Conference, 2003:86–90.

30. Eastwood JW. The virtual particle electromagnetic particle-mesh method. *Computer Physics Communication,* 1991, 64(2):252–266.

31. Pointon TD, Desjarlais MP. Three-dimensional, particle-in-cell simulations of applied-B ion diodes on the particle beam fusion accelerator II[J]. *Journal of Applied Physics,* 1996, 80(4):2079–2093.

32. Cao LH, Liu DQ, Chang WW, et al. Multi-time-scale algorithm for two-dimensional particle simulation code. *Journal of National University of Defense Technology,* 1996, 18(3):133–137 (in Chinese).

33. Mo ZY, Xu LB, Zhang LB, et al. Parallel computing and performance analysis for 2-dimensional plasma simulations with particle clouds in cells methods. *Chinese Journal of Computational Physics,* 1999, 16(5):496–504 (in Chinese).

34. Ma YY, Chang WW, Yin Y, et al. A collision model in plasma particle simulations. *Acta Physica Sinica,* 2000, 49(8):1513–1519 (in Chinese).

35. Yin Y, Chang WW, Ma YY, et al. Gridless particle simulation of electrostatic plasmas. *Chinese Journal of Computational Physics,* 2001, 18(3):263–270 (in Chinese).

36. Liu DQ, Cao LH, Chang WW, et al. Review on a 2-dimensional particle simulation program of laser plasma interaction. *Chinese Journal of Computational Physics,* 1998, 15(6):641–647 (in Chinese).

37. Zhuo HB, Chang WW, Zhuo HC. The object-orient plasma simulation code in cylindrical coordinates. *Journal of National University of Defense Technology,* 2001, 23(2):103–106 (in Chinese).

38. Xu H, Chang WW, Zhuo HB, et al. Parallel programming of 2 1/2-dimensional PIC under distributed-memory parallel environments. *Chinese Journal of Computational Physics,* 2002, 19(4):305–310 (in Chinese).

39. Ma YY, Chang WW, Yin Y, et al. 2 1/2-dimensional particle simulation code of laser-plasma. *Chinese Journal of Computational Physics,* 2002, 19(4):311–316 (in Chinese).

40. Ma YY, Chang WW, Yin Y, et al. Three dimensional object oriented parallel particle simulation Code-PLASIM3D. *Chinese Journal of Computational Physics,* 2004, 21(3):305–311 (in Chinese).

41. Chen H, Zhang X, Xia F, et al. A data model and I/O performance improvement for 3-D plasma simulations with particle. *Computer Engineering and Applications,* 2004, 20:104–107 (in Chinese).

42. Liu DG, Zhu DJ, Zhou J, et al. Design of 3D particle-in-cell simulation software. *High Power Laser and Particle Beams*, 2006, 18(1):110–114 (in Chinese).
43. Zhou J, Liu DG, Liu SG. Object-oriented design of CAD system for PIC simulation. *Acta Electronica Sinica*, 2008, 36(3):556–561 (in Chinese).
44. Li YD, Zhang DH, Liu CL, et al. 2.5-dimensional electromagnetic PIC general simulation software for high power microwave devices-UNIPIC, Military Microwave Tube Symposium at the 15th Annual Academic Conference, Kunming, 2005, 475–479(in Chinese).
45. Li YD, He F, Liu CL, et al. Improved electromagnetic particle-in-cell algorithm for plasma simulation. *Journal of Xi'an Jiaotong University*, 2002, 36(10):1000–1003(in Chinese).
46. Yang WY, Dong Y, Chen J, et al. Simulation and comparison of MILO by parallel 3D full electromagnetic particle simulation software, The 7th National High Power Microwave Symposium, 2008: 187–194(in Chinese).
47. Dong Y, Chen J, Yang WY, et al. Massively parallel code named NEPTUNE for 3D fully electromagnetic and PIC simulations. *High Power Laser and Particle Beams*, 2011, 23(6):1607–1615 (in Chinese).
48. Chen ZG, Wang Y, Wang JG, et al. Parallelization techniques in 3D parallelized fully-electromagnetic particle simulation code. *Computer Engineering Science*, 2009, 31(11):128–131 (in Chinese).
49. Liao C, Zhu DJ, Liu SG. Design of parallel algorithm for Poisson module of CHIPIC software. *Journal of University of Electronic Science and Technology of China*, 2008, 37(1):81–83 (in Chinese).
50. Liao C, Liu DG, Liu SG. Tree-dimensional electromagnetic particle-in-cell simulation by parallel computing. *Acta Physica Sinica*, 2009, 58(10):6709–6718 (in Chinese).
51. Shao FQ. *Plasma Particle Simulation*. Beijing: Science Press, 2002 (in Chinese).
52. Verboncoeur JP. Particle simulation of plasmas: review and advances. *Plasma Physics and Controlled Fusion*, 2005, 47(4):A231–A260.
53. Zhou J. *Research on Electromagnetic Particle Simulation Method and Its Application*. Chengdu: University of Electronic Science and Technology, 2009 (in Chinese).
54. Liao C. *Application and Research of 3D Electromagnetic Particle Parallel Algorithm*.Chengdu: University of Electronic Science and Technology, 2010 (in Chinese).
55. Okuda H, Birdsall CK. Collisions in a plasma of finite-size particles. *Physics of Fluids*, 1970, 13(8):21–23.
56. Boris JP. Relativistic plasma simulation—optimization of a hybrid code, Proceedings on 4th Conference on Numerical Simulation of Plasmas. Washington, DC, 1970:3–67.
57. Cheng H, Jiang JP. *Cathode Electronics*. Xi'an: Northwest Telecommunications Engineering Institute Press, 1986 (in Chinese).
58. Liu XQ. *Cathode Electronics*. Beijing: Science Press, 1980 (in Chinese).

59. Jiang JP, Weng JH, Yang PT, et al. *Cathode Electronics and Principles of Gas Discharge*. Beijing: National Defense Industry Press, 1980 (in Chinese).
60. Nelson EM. An analysis of the basic space-charge-limited emission algorithm in a finite-element electrostatic gun code. *IEEE Transactions on Plasma Science*. 2004, 32(3):1223–1234.
61. Pandey BK, Kulshrestha S, Dey R, et al. Square coaxial beam forming network for multilayer microstrip antenna, Proceedings of Asia-Pacific Microwave Conference, 2007.
62. Katzenmeyer AM, Leonard F, Talin AA, et al. Observation of space-charge-limited transport in InAs Nanowires. *IEEE Transactions on Nanotechnology*, 2011, 10(1):92–95.
63. Yeh YS, Tsao MH, Chen HY, et al. Improved computer program for magnetron injection gun design. *International Journal of Infrared and Millimeter Waves*, 2000, 21(9):1397–1415.
64. Lawson W. Space-charge limited magnetron injection guns for high-power gyrotrons. *IEEE Transactions on Plasma Science*, 2004, 32(3):1236–1241.
65. Uimanov, IV. PIC simulation of the electric field at a cathode with a surface micro protrusion under intense field emission, 23rd International Symposium on Discharges and Electrical Insulation in Vacuum, 2008.
66. Barker RJ, Schamiloglu E. *High-Power Microwave Sources and Technologies*. New York: John Wiley & Sons, Inc., 2001.
67. Shi L, Zhang JS, Qiu AC, et al. Initial energy of electrons effect on space-charge-limited current densities in two-component fluxe diode. *High Power Laser and Particle Beams*, 2001, 13(3):357–359 (in Chinese).
68. Song SY, Chou X, Wang WD, et al. Dependence of Space-charge-Limited flow on conducting current. *High Power Laser and Particle Beams*, 2005, 17(3):441–446 (in Chinese).
69. Liu GZ. Modification to space-charge-limited current of intense electron beam diode. *High Power Laser and Particle Beams*, 2000, 12(3):375–378(in Chinese).
70. Ge DB, Yan YB. *Electromagnetic Wave Finite Difference Time Domain Method (3rd edition)*. Xi'an: Xidian University Press, 2011 (in Chinese).
71. Wang BZ. *Computational Electromagnetics*. Beijing: Science Press, 2002 (in Chinese).
72. Berenger JP. A perfectly matched layer for the absorption of electromagnetic waves. *Journal of Computational Physics*, 1994, 114(2):185–200.
73. Berenger JP. Three-dimensional perfectly matched layer for the absorption of electromagnetic waves. *Journal of Computational Physics*, 1996, 127(2):363–379.
74. Gedney SD. An anisotropic perfectly matched layer absorbing media for the truncation of FDTD lattices. *IEEE Transactions on Antennas and Propagation*, 1996, 44(12):1630–1639.

CHAPTER 4

EM PIC Simulation of Multipactor

4.1 INTRODUCTION

Under vacuum conditions, when the input power of the microwave components exceeds a certain threshold, the electrons accelerated by the electromagnetic field could excite secondary electron emission (SEE) between the two metal surfaces to form a resonant electron multiplication effect, which eventually leads to multipactor. Multipactor generated in this process is a nonlinear and strongly coupled physical phenomenon, and its analysis and modelling involve cross-fusion of multiple disciplines. The traditional analytical method and linear analysis alone cannot accurately predict multipactor and the discharge breakdown threshold power, so the numerical method is required. This chapter explores three-dimensional numerical simulation techniques for the motion of charged particles in the electromagnetic field and the interaction between charged particles and the electromagnetic field based on particle simulation techniques. Through the time-sharing linear numerical solution of the Maxwell equation that regulates the evolution of the electromagnetic field and the Lorentz equation that regulates the motion of charged particles to simulate the self-consistent nonlinear coupling effect between both, finally giving real-time information on the evolution of the electromagnetic field and the motion of the charged particles and realizing the numerical simulation of the SE multiplication effect. In this chapter, combining with the

DOI: 10.1201/9781003189794-4

theory of SEE, the three-dimensional numerical simulation technique of multipactor is introduced and specified, and the case studies of multipactor numerical simulations for typical spacecraft microwave components are also illustrated on the basis of the electromagnetic calculation and PIC simulation method introduced in the previous chapter.

4.2 EM PIC SIMULATION METHOD

In theory, the core algorithm of the finite difference time domain (FDTD) [1–4] is relatively independent from that of the particle-in-cell (PIC) [5–11], and they can work independently in the case that field and particle coupling are not involved. For the conventional FDTD method, the spatial discretization was made based on a finite difference Yee lattice. Without involving the nonlinear interaction effect of the field and charged particles, the time step only needs to meet the CFL stability condition. After introducing the corresponding boundary conditions, the problem can be solved using mathematical models. The conventional PIC could adopt any meshing method for space discretization and is without specific stability condition. It focuses itself on the particle emission model and the solving scale problem.

Multipactor in the microwave components of the spacecraft is a direct result of the strong nonlinear interaction between the electromagnetic field and the particles[12–25], which cannot accurately be simulated by the simple linear superposition of FDTD and PIC. Multipactor is a complex nonlinear process, which originates from the SEE phenomenon of materials. Since the complexity of the actual electromagnetic field distribution and the randomness of the electron motion, it is almost impossible to use analytical methods to study multipactor in practical microwave components. Therefore, in this chapter, we discuss the use of first-principles numerical calculation methods to simulate the complete multipactor process and predict the breakdown threshold power in multipactor. Based on first-principles, the electrons will remain in motion until the conditions for the end of the operation are satisfied, or until the end of the total calculation time. At the start, the initial conditions of electrons and fields should be defined and a proper expression of the fundamental physical laws of electron motion and electromagnetic field evolution (such as Maxwell equations and Newton-Lorentz motion equations) should be given. The electromagnetic field distribution in a given structure is obtained by solving Maxwell difference equations in discrete meshes by the FDTD method. The evolution of the electron motion is obtained by solving the difference form of the relative Lorentz motion equation in the time domain. The charges and currents generated by the electron motion are interpolated at each node of the discrete mesh as the source terms. By recalculating the updated electromagnetic field at each time step, taking the space charge effect into account, the interaction process of electromagnetic field-electron and electron-electromagnetic field is simulated.

To simulate the process of particle-field interaction, the Maxwell equations solved by the FDTD method and the Newton-Lorentz equation solved by the PIC method are coupled by the current generated by the electron motion. First, the Maxwell equation is discrete in the Yee lattice (as shown in Figure 4.1), and the electromagnetic field calculation is carried out by the FDTD method; Newton-Lorentz equation is solved by Boris "half-acceleration-rotation-half-acceleration" method; the field equations and the equations of particles motion are coupled by the current density term of the Maxwell equation as follows: the field at any space position of the particles can be obtained by the interpolation of the field at the discrete lattice point, and the source terms ρ and J (charge density and current density) in the field equation are interpolated.

As shown in Figure 4.2, the flow of the co-simulation algorithm for multipactor electrons is that the particle speed at the time of $n-\dfrac{1}{2}$ is advanced to the time of $n+\dfrac{1}{2}$, and the particle position at the time of n

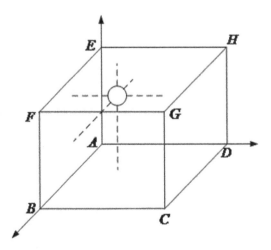

FIGURE 4.1 Electron motion in the three-dimensional Yee lattice.

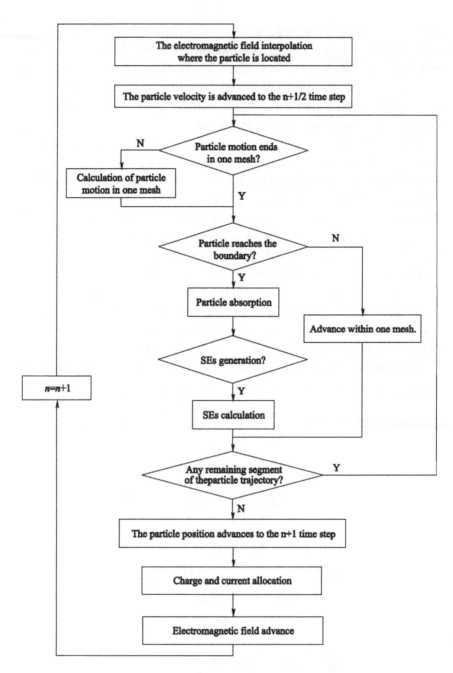

FIGURE 4.2 Cycle iteration flow of multipactor simulation algorithm.

is advanced to the time of $n+1$. The staggered particle motion algorithm is established under the consideration of the boundary effect on particles, including absorption, reflection and SEE.

4.3 EM PIC SIMULATION METHOD BASED ON NON-UNIFORM MESHING

For the microwave devices with complex structure, some parts have small characteristic scale, while some parts have large size. In order to take both of the simulation accuracy and the calculation efficiency into account, a non-uniform meshing method is needed. A fine meshing method is used where the scale is small and has a great impact on the device performance, and a coarse meshing method is used where the scale is large and the electromagnetic field changes little.

In the three-dimensional simulation of particles in the non-uniform meshes, the particle location is transformed from the real space to the mesh space, that is, the particle location is represented by the relative coordinates in meshes.

The mesh coordinates x of particles contain integer and decimal parts, which can be represented as $(i+\alpha, j+\beta, k+\gamma)$, where i, j, k are integer numbers and α, β, γ are fraction numbers $(\alpha, \beta, \gamma) \in [0,1)$. The integer part is the number of the mesh, and the decimal part is the relative position of the particle in the mesh. By the mesh coordinate representation method, the actual coordinates should be converted into the mesh coordinate in a non-uniform mesh during the calculation process.

4.3.1 Mesh Coordinates Converted to Actual Coordinates

The actual coordinates corresponding to the mesh points are represented by a three-dimensional array $\mathbf{G}$, where $\mathbf{G}_{i,j,k} = (G_1, G_2, G_3)_{i,j,k}$ is a vector and represents the actual coordinates in all directions. If the mesh coordinate $(i_q, j_q, k_q) = (i+\alpha, j+\beta, k+\gamma)$ of a certain point q is shown in Figure 4.3, its actual coordinate $x\_MKS = (x_{1q}, x_{2q}, x_{3q})$ can be linearly interpolated from the actual coordinate G of the surrounding mesh points:

$$x\_MKS = (1-\alpha)(1-\beta)(1-\gamma)G_{i,j,k} + (1-\alpha)(1-\beta)\gamma G_{i,j,k+1}$$

$$+(1-\alpha)\beta(1-\gamma)G_{i,j+1,k} + (1-\alpha)\beta\gamma G_{i,j+1,k+1}$$

$$+\alpha(1-\beta)(1-\gamma)G_{i+1,j,k} + \alpha(1-\beta)\gamma G_{i+1,j,k+1}$$

$$+\alpha\beta(1-\gamma)G_{i+1,j+1,k} + \alpha\beta\gamma G_{i+1,j+1,k+1} \qquad (4.1)$$

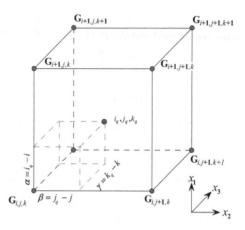

FIGURE 4.3 Convert mesh coordinates to actual coordinates.

4.3.2 Actual Coordinates Converted to Mesh Coordinates

To convert the actual coordinates to mesh coordinates, it is necessary to determine which mesh cell the particle is located in. In uniform meshes, the coordinates can be calculated through classical equations. In in nonuniform meshes, it is necessary to compare the actual coordinates with mesh coordinates one by one. For point q with the actual coordinate of $x\_MKS=(x_{1q}, x_{2q}, x_{3q})$, if

$$G_{1\,i,0,0} \leq x_{1q} < G_{1\,i+1,0,0}$$

$$G_{2\,0,j,0} \leq x_{2q} < G_{2\,0,j+1,0} \qquad (4.2)$$

$$G_{3\,0,0,k} \leq x_{3q} < G_{3\,0,0,k+1}$$

then point q is in element (i, j, k), that is, the integral part of the mesh coordinate of point q. The decimal part of the mesh coordinate of point q is calculated as follows:

$$\alpha = \frac{x_{1q} - G_{1\,i,j,k}}{G_{1\,i+1,j+1,k+1} - G_{1\,i,j,k}} = \frac{x_{1q} - G_{1\,i,j,k}}{\Delta G_{1\,i,j,k}}$$

$$\beta = \frac{x_{2q} - G_{2\,i,j,k}}{G_{2\,i+1,j+1,k+1} - G_{2\,i,j,k}} = \frac{x_{2q} - G_{2\,i,j,k}}{\Delta G_{2\,i,j,k}} \qquad (4.3)$$

$$\gamma = \frac{x_{3q} - G_{3\,i,j,k}}{G_{3\,i+1,j+1,k+1} - G_{3\,i,j,k}} = \frac{x_{3q} - G_{3\,i,j,k}}{\Delta G_{3\,i,j,k}}$$

4.4 BOUNDARY CONDITIONS IN MULTIPACTOR SIMULATION

When the FDTD method is used to simulate the electromagnetic field distribution, the boundary conditions should be set before the simulation. The general boundary types include Perfect Electrical Conductor (PEC) boundary (electrical wall boundary), lossy conductor boundary, open boundary (including electromagnetic wave injection and absorption boundary), Perfect Magnetic Conductor (PMC) boundary, symmetric boundary, periodic boundary and so on. For particle propulsion calculation, the boundary types mainly include particle absorption boundary, particle emission boundary, symmetric boundary, periodic boundary and so on. In co-simulations of FDTD and PIC, if the electromagnetic field boundary and particle boundary are set separately, it is cumbersome and not customary. As in the practical microwave device, the metal conductor is the electric wall boundary of the electromagnetic field FDTD calculation and also the SEE boundary of the PIC calculation; the open boundary absorbs electromagnetic waves while also absorbing particles. The particle-field boundaries are therefore common in practice, and common boundary processing is also required in the co-simulations of FDTD and PIC, which are described separately below.

4.4.1 Conductor Boundary

For EM calculation, the conductivity of the real conductor is not infinite, and the current conducted in the skin depth will cause power losses. Due to the existence of the tangential electric field on the conductor surface, the loss needs to be considered when modelling, that is, the lossy conductor model. If the conductor losses are very small over a certain frequency range, it can generally be considered as an ideal conductor, i.e. an electrical wall boundary with a tangential electric field of zero.

For particle motion calculations, when electrons are incident to the conductor boundary, they will interact with the material, part of which will be elastic or non-elastic reflections, and part of which will excite a SE. This is also the process of SEE. The processing algorithm for the SE generation boundary will be described in detail in subsequent sections.

4.4.2 Open Boundary

In order to improve calculation efficiency and reduce memory overhead, the calculation area needs to be truncated at certain positions. This artificial truncation occurs where an open boundary needs to be added to ensure the correctness of the calculation results.

For EM calculation, the open boundaries include the electromagnetic wave injection boundary and the electromagnetic wave absorption boundary.

For particle calculations, open boundaries also include particle absorption and particle injection (emission) boundaries. The former is more commonly used, and few particles are injected at the open boundary.

Under the perfectly matched layers (PMLs) as the electromagnetic absorption boundary [13], when the particles pass through this type of boundary, these particles could not be directly deleted from the calculation area without any other processing. Then, the charge in the calculation region will not be conserved, resulting in errors. The PML absorption boundary based on the principle of attenuation of electromagnetic waves by artificial media is more dependent on the field equation. Thus, it is greatly affected by particles passing through the boundary. This section proposes a correction method to eliminate the effect of particles passing through the PML boundary. Similar to the B. Marder's Langdon-Marder error correction method, the divergence error is defined as:

$$\delta\rho = \nabla \cdot \mathbf{D} - \rho \qquad (4.4)$$

Diffuse the divergence to the surrounding calculation region, i.e.

$$\varepsilon \frac{E_{\text{corrected}}^{n+1} - E^{n+1}}{\Delta t} + \frac{\sigma E_{\text{corrected}}^{n+1} + \sigma E^{n+1}}{2} = \nabla[d(\delta\rho^n)] \qquad (4.5)$$

The value range of the diffusion coefficient d is the same as in Langdon-Marder correction. Absorb particles passing through electromagnetic boundaries and reduce the impact of particle disappearance on calculations. Thus, an open boundary shared by electromagnetic particles is constructed.

4.4.3 Symmetry Boundary

For electromagnetic field calculations, the symmetry boundaries include electric field symmetry and magnetic field symmetry, i.e., zero normal electric field and zero normal magnetic field, respectively.

For particle calculations, particles incident to the symmetric boundary are reflected back to the calculation region. That is, the velocity of the incident particle changes direction in the direction of the normal vector of the boundary without changing size, and the velocity components of the other two directions do not change the direction and size.

4.4.4 Periodic Boundary

Periodic boundary refers to the existence of periodicity in a certain direction, which occurs in pairs, i.e. in the position of the maximum and minimum values in a certain direction.

For EM calculations, the field outside the periodic boundary is equal to the field inside the periodic boundary on the other side.

For particle calculations, particles incident to the periodic boundary will enter the calculation area from the periodic boundary on the other side with the unchanged magnitude and direction of velocity.

4.5 EFFECT OF SEE ON MULTIPACTOR SIMULATION

4.5.1 Basic Theory

There are two important parameters for the description of SEE: the secondary electron yield (SEY) δ and the emission energy spectrum ($\frac{d\delta}{dE}$). The SEY could be defined as follows:

$$\delta = \frac{I_s}{I_0}, \tag{4.6}$$

where I_0 is the incident electron beam current and I_s is the current formed by SEs emitted from the material surface. The SEY is a function of the incident electron energy, incident angle and surface properties.

By adding a blocking voltage V to the SE current detector, the SE current with the value of $E_k > E$ can be detected. The SEE spectrum is defined as:

$$S(E_0, E) = \frac{I_s(E)}{I_0} \tag{4.7}$$

Consider that $S(E_0,E)$ depends not only on the emission energy but also on the incident angle θ_0, $\dfrac{d\delta}{dE}$ is defined as follows:

$$\frac{d\delta}{dE} = -\frac{\partial S(E_0,E)}{\partial E} \tag{4.8}$$

The minus sign is to guarantee $\dfrac{d\delta}{dE} > 0$. $I_s(E)$ is the SE current that overcomes the blocking voltage V, namely $S(E_0,0) = \delta(E_0)$, when $E > E_0$, $I_s(E) = 0$, then:

$$\int_0^\infty dE \frac{d\delta}{dE} = \delta(E_0) \tag{4.9}$$

In order to describe a specific SEE process, more variables or measurement parameters need to be considered. For example, an additional azimuth angle ϕ_0 (different from the polar angle θ_0) should be considered for the anisotropic material. The difference is caused by the material surface manufacturing process, the crystal axial properties or grooves. If the SE detector can determine the emission angle of the SEs, an additional emission angle should be considered in S; if the incident electron beam is polarized, SEY and the emission energy spectrum need to be defined separately for the different spin polarization modes of the incident electrons.

Assuming that an electron with energy E_0 collides with the surface of the material at an angle of θ_0 and emits n SEs with probability $P_n (n = 1...\infty)$, it is clear that P_n meets the following conditions.

$$\sum_{n=0}^\infty P_n = 1, P_n \geq 0 \tag{4.10}$$

where P_0 is the probability of the incident electron being absorbed. The expected value of SEY can be expressed as $\delta = <n> = \sum_{n=1}^\infty nP_n$.

Taking into account additional SEE information, the phase space probability is defined as:

$$\mathbb{P}_n = \frac{dP_n}{dE_1 d\Omega_1 dE_2 d\Omega_2 ... dE_n d\Omega_n} \tag{4.11}$$

where E is the energy and Ω is the spatial angle $[\Omega_n = (\theta_n, \phi_n)]$ of the SE. According to the definition of phase space probability $\mathbb{P}_n$, the probability of emitting n SEs at this time can be obtained:

$$P_n = \int_n (dE) \, (d\Omega)_n \mathbb{P}_n, \quad n \geq 1 \tag{4.12}$$

The absorption probability of SEs is:

$$P_0 = 1 - \sum_{n=1}^{\infty} P_n \tag{4.13}$$

Assuming the energy of the incident electron is E_0, the energy spectrum of this electron is:

$$S(E_0, E) = \sum_{n=1}^{\infty} \int (dE)_n (d\Omega)_n \mathbb{P}_n \sum_{k=1}^{n} \theta(E_k - E) \tag{4.14}$$

where $\theta(E_k - E)$ is used to guarantee that only the probability of $E_k > E$ is calculated. According to the above formula, the expression of $\dfrac{d\delta}{dE}$ can be written equivalently according to formula (4.8) as follows:

$$\frac{d\delta}{dE} = \sum_{n=1}^{\infty} \int (dE)_n (d\Omega)_n \mathbb{P}_n \sum_{k=1}^{n} \delta(E_k - E) \tag{4.15}$$

4.5.2 Numerical SEE Model

4.5.2.1 Emission Angle

For simplicity, the function of $\mathbb{P}_n$ is expressed as follows:

$$\mathbb{P}_n = A_n(\Omega_1, \ldots, \Omega_n) \times \frac{dP_n}{(dE)_n} \tag{4.16}$$

The emission angle of the true secondary electron (TSE) conforms to the cosine distribution $\cos\theta$, while the elastic reflection and scattering have a more complex scattering angle distribution. For the convenience of analysis, it is assumed that all SEs conform to cosine distribution:

$$A_n\left(\Omega_1,\ldots,\Omega_n\right)=\left(\frac{\alpha+1}{2\pi}\right)^n\cos^\alpha\theta_1\cos^\alpha\theta_2\ldots\cos^\alpha\theta_n \qquad (4.17),$$

where α is an adjustable parameter approaching 1. Obviously:

$$\int_n\left(d\Omega\right)\ A_n\left(\Omega_1,\ldots,\Omega_n\right)=1 \qquad (4.18)$$

The distribution of the energy can be expressed as:

$$\frac{dP_n}{(dE)_n}=\int\left(d\Omega\right)_n\mathbb{P}_n \qquad (4.19)$$

4.5.2.2 Emission Energy

Suppose $\dfrac{dP_n}{(dE)_n}$ has the following form:

$$\frac{dP_n}{(dE)_n}=\theta\left(E_0-\sum_{k=1}^{n}E_k\right)\prod_{k=1}^{n}f_n\left(E_k\right)\theta\left(E_0-E_k\right) \qquad (4.20),$$

where $f_n\left(E_k\right)$ is the energy distribution of the kth SE among n SEs. θ function is used to guarantee that the total energy of the SEs does not exceed the incident energy. $\theta\left(E_0-E_k\right)$ is used to guarantee that the energy of each SE does not exceed the energy of the incident electron.

4.5.3 Three Types of SEs

4.5.3.1 Basic Assumptions

When a stable incident current I_0 collides on the surface of the material, part of the current I_e is caused by elastic reflection; the rest enters into the material. If the electrons that enter the interior of the material emit out of the surface after collision, they are called scattering electrons, this part of the current is represented as I_r. There are also some electrons that have complex interaction mechanism with the material, which are called TSE reflection, this part of the current is represented as I_{ts}. The yield of each type of secondary electrons is defined as follows: $\delta_e=\dfrac{I_e}{I_0}$, $\delta_r=\dfrac{I_r}{I_0}$, and $\delta_{ts}=\dfrac{I_{ts}}{I_0}$. The total SEY can be written as:

$$\delta = \frac{(I_e + I_r + I_{ts})}{I_0} = \delta_e + \delta_r + \delta_{ts} = P_1 + 2P_2 + 3P_3 + \cdots \quad (4.21)$$

In order to calculate the responding probability P from the SEY δ of these three parts of SEs, the assumptions are made that only one electron ($n = 1$) is produced by the back reflection electron. Any quantities of SEs ($n \geq 1$) are produced by the TSE. According to the assumption, P_n can be expressed as follows:

$$P_1 = P_{1,e} + P_{1,r} + P_{1,ts}$$
$$P_n = P_{n,ts} \quad n \geq 2$$

$$(4.22)$$

The absorption probability is:

$$P_0 = 1 - \sum_{n=1}^{\infty} P_n = 1 - P_{1,e} - P_{1,r} - \sum_{n=1}^{\infty} P_{n,ts} \quad (4.23),$$

and hence

$$\delta_e = P_{1,e}$$
$$\delta_r = P_{1,r} \quad (4.24)$$
$$\delta_{ts} = \sum_{n=1}^{\infty} nP_{n,ts}.$$

The energy distribution f_n can be written as:

$$f_1 = f_{1,e} + f_{1,r} + f_{1,ts}$$
$$f_n = f_{n,ts} \quad n \geq 2$$

$$(4.25)$$

All of the above $f_{1,e}, f_{1,r}, f_{n,ts}, P_{1,e}, P_{1,r}$ and $P_{n,ts}$ must be defined before it can be used in calculation.

4.5.3.2 Elastic Electron Model

Obviously, the equation of $\delta_e(E_0, \theta_0)$ at ($\theta_0 = 0$) obtained from the experimental data can be expressed as follows:

$$\delta_e(E_0,0) = P_{1,e}(\infty) + \left(\hat{P}_{1,e} - P_{1,e}(\infty)\right)e^{-\left(|E_0 - \hat{E}_e|/w\right)^p/p} \tag{4.26}$$

Assuming that $\hat{P}_{1,e} > P_{1,e}(\infty)$ satisfies the actual situation, the function reaches its peak value at $E_0 = \hat{E}_e$. For the energy probability function $f_{1,e}$, the form that roughly satisfies the elastic reflection part of $\dfrac{d\delta}{dE}$ can be given as:

$$f_{1,e} = \theta(E)\theta(E_0 - E)\delta_e(E_0,\theta_0)\frac{2e^{-(E-E_0)^2/2\sigma_e^2}}{\sqrt{2\pi}\sigma_e\,\mathrm{erf}\left(E_0/\sqrt{2}\sigma_e\right)} \tag{4.27}$$

$f_{1,e}$ has been normalized and therefore satisfies:

$$\int_0^{E_0} dE f_{1,e}(E) = \delta_e(E_0) \tag{4.28}$$

In fact, $\delta_e \neq 0$ when $E_0 = 0$, which is different from equation (4.26). This difference is not caused by physical factors, but $f_{1,e}$ is integrated according to equation (4.27).

4.5.3.3 Scattering Electron Model

According to the experimental data, the expression of $\delta_e(E_0,\theta_0)$ at $(\theta_0 = 0)$ can be obtained as follows:

$$\delta_r(E_0,0) = P_{1,r}(\infty)\left[1 - e^{-(E_0/E_r)^r}\right] \tag{4.29}$$

Suppose $f_{1,r}$ is:

$$f_{1,e} = \theta(E)\theta(E_0 - E)\delta_e(E_0,\theta_0)\frac{(q+1)E^q}{E_0^{q+1}} \tag{4.30}$$

$f_{1,e}$ has been normalized and therefore satisfies:

$$\int_0^{E_0} dE f_{1,r}(E) = \delta_r(E_0) \tag{4.31}$$

4.5.3.4 True Secondary Electrons

Through an approximate scaling function $D(x)$, it can be ensured that $\delta_{ts}(E,\theta)$ is in good accordance with the experimental data:

$$D(x)=\frac{sx}{s-1+x^s} \tag{4.32}$$

where s is an adjustment function greater than 1. It can be assumed that the energy spectrum distribution function of TSEs is as follows

$$f_{n,ts}=\theta(E)F_nE^{p_n-1}e^{-\frac{E}{\varepsilon_n}}, \tag{4.33}$$

where p_n and ε_n are phenomenal parameters, which have no practical physical significance.

$$F_n^n=\frac{P_{n,ts}\left(E_0\right)}{\left(\varepsilon_n^{p_n}\Gamma(p_n)\right)^n P\left(np_n,\frac{E_0}{\varepsilon_n}\right)} \tag{4.34}$$

$P(z,x)$ is the standard incomplete Gamma function. The relationship between emission electron spectrum probabilities $P_{n,ts}$ and δ_{ts} will be given in the next section. Although equation (4.33) does not conform to the expression form proposed by the theory of SEE of metals, it can well conform to most of the experimental data. Moreover, this expression is good for integration and very suitable for phenomenological analysis.

4.5.3.5 Emission Probability

The relationship between $P_{n,ts}$ and δ_{ts} can be obtained from formulae (4.22) to (4.24) as follows:

$$P_0 = P_{0,ts}-\delta_e-\delta_r$$
$$P_1 = P_{1,ts}+\delta_e+\delta_r \tag{4.35}$$
$$P_n = P_{n,ts}, \quad n\geq 2$$

Since $P_{n,ts}$ satisfies the normalization regulation, then

$$\sum_{n=0}^{\infty} P_{n,ts} = 1 \tag{4.36}$$

P_n can be defined to satisfy the Poisson distribution:

$$P_{n,ts} = \frac{\delta_{ts}^n}{n!} e^{-\delta_{ts}}, 0 \leq n < \infty \tag{4.37}$$

$P_{n,ts}$ meets the expected value $\langle n \rangle = \delta_{ts}$. It can also be defined that $P_{n,ts}$ satisfies binomial distribution:

$$P_{n,ts} = \begin{pmatrix} M \\ n \end{pmatrix} p^n (1-p)^{M-n}, 0 \leq n \leq M, \tag{4.38}$$

where $p = \dfrac{\langle n \rangle}{M} = \dfrac{\delta_{ts}}{M}$. This distribution has a limiting parameter M related to the maximum of SEY. Experiments show that when $M = 10$, it can make a good fit accuracy while simulating the SEE characteristics of most materials.

4.5.3.6 Correction of TSE Emission Probability

Although $\delta_e + \delta_r \leq 1$ can be guaranteed, P_0 in equation (4.35) may exceed 1, and P_1 in equation (4.35) may still be negative. The above-mentioned situation will occur when $\delta_{ts} \geq 1.2$ and $\delta_e + \delta_r \geq 0.5$. When calculating the yield of TSEs, the total incident electron current is calculated to ensure that $P_0 \geq 0$ and $P_1 \leq 1$. That is, using $I_0 - I_e - I_r$ instead of the total current in the yield of TSE:

$$\delta'_{ts} = \frac{I_{ts}}{I_0 - I_e - I_r} = \frac{\delta_{ts}}{1 - \delta_e - \delta_r} \tag{4.39}$$

Expressed by probability as

$$\delta'_{ts} = \sum_{n=1}^{\infty} n P'_{n,ts} \tag{4.40}$$

According to equation (4.24), then:

$$P_{n,ts} = \frac{\delta_{ts}}{\delta_{ts}'} P_{n,ts}' = \left(1 - \delta_e - \delta_r\right) P_{n,ts}', \quad n \geq 1 \tag{4.41}$$

According to equations (4.23) to (4.25), the emission probability of n SEs can be obtained for each incident electron

$$P_0 = \left(1 - \delta_e - \delta_r\right) P_{0,ts}',$$

$$P_1 = \left(1 - \delta_e - \delta_r\right) P_{1,ts}' + \delta_e + \delta_r, \tag{4.42}$$

$$P_n = \left(1 - \delta_e - \delta_r\right) P_{n,ts}', \quad n \geq 2,$$

The following equation shall be met:

$$\sum_{n=0}^{\infty} P_{n,ts} = 1 - \delta_e - \delta_r \tag{4.43}$$

Assuming that P_n is subjected to Poisson distribution:

$$P_{n,ts}' = \frac{\delta_{ts}'^n}{n!} e^{-\delta_{ts}'}, 0 \leq n < \infty \tag{4.44}$$

$P_{n,ts}'$ meets the expected value $\langle n \rangle = \delta_{ts}'$.
    So assumed that $P_{n,ts}'$ is in the binomial distribution:

$$P_{n,ts}' = \left(\begin{array}{c} M \\ n \end{array} \right) p^n (1-p)^{M-n}, 0 \leq n \leq M \tag{4.45}$$

4.5.3.7 Relationship with the Incident Angle
Furman mentioned in his study that it has been proved by experiments in a vacuum pressure chamber that when the incident angle is $0 \leq \theta_0 \leq 84°$, it can be well connected with the incident angle by the total SEY multiplied by a factor in the form of $1 + \alpha_1 \left(1 - \cos^{a_0} \theta_0\right)$. Assuming that the three components of elastic electron, scattering electron and TSE in the SEY are all

suitable for this form, the elastic electron and scattering electron part can be represented as:

$$\delta_e(E_0,\theta_0) = \delta_e(E_0,0) \times \left[1 + e_1\left(1 - \cos^{e_2}\theta_0\right)\right]$$

$$\delta_r(E_0,\theta_0) = \delta_r(E_0,0) \times \left[1 + r_1\left(1 - \cos^{r_2}\theta_0\right)\right]$$

$$(4.46)$$

The expression of the TSE part is as follows:

$$\hat{\delta}(\theta_0) = \hat{\delta}_{ts} \times \left[1 + t_1\left(1 - \cos^{t_2}\theta_0\right)\right]$$

$$\hat{E}(\theta_0) = \hat{E}_{ts} \times \left[1 + t_3\left(1 - \cos^{t_4}\theta_0\right)\right]$$

$$(4.47)$$

4.5.3.8 Secondary Electron Emission Spectrum

In order to extract more information from the data, the formula of energy spectrum should be established. According to the above expression and the model of $\mathbb{P}_n$, then:

$$\frac{d\delta}{dE} = \sum_{n=1}^{\infty} n f_n(E) \int_0^{\infty} \prod_{k=2}^{n} \{dE_k f_n(E_k)\} \theta\left(E_0 - E - \sum_{k=2}^{n} E_k\right)$$

$$= \sum_{n=1}^{\infty} n f_n(E) P_{n-1}(E_0 - E)$$

$$(4.48)$$

The above formula conforms to the summation rule:

$$\int_0^{E_0} dE \frac{d\delta}{dE} = \sum_{n=1}^{\infty} n P_n = \delta(E_0)$$

$$(4.49)$$

According to the elastic electron, scattering electron and TSE in SEY described above, the total energy spectrum can be obtained as follows:

$$\frac{d\delta}{dE} = f_{1,e} + f_{1,r} + \frac{d\delta_{ts}}{dE},$$

$$(4.50)$$

where

$$\frac{d\delta_{ts}}{dE} = \sum_{n=1}^{\infty} \frac{nP_{n,ts}(E_0)\left(\dfrac{E}{\varepsilon_n}\right)^{p_n-1} e^{-E/\varepsilon_n}}{\varepsilon_n \Gamma(p_n) P\left(np_n, \dfrac{E_0}{\varepsilon_n}\right)} P\left[(n-1)p_n, \frac{(E_0-E)}{\varepsilon_n}\right], \quad (4.51)$$

where $P(z,x)$ is the standard incomplete gamma function, satisfying $P(0,x)=1$. When $E_0 \gg E, \varepsilon_n$, a simplified expression can be obtained:

$$\frac{d\delta_{ts}}{dE} = \sum_{n=1}^{\infty} nP_{n,ts}(E_0) \frac{\left(\dfrac{E}{\varepsilon_n}\right)^{p_n-1} e^{-\dfrac{E}{\varepsilon_n}}}{\varepsilon_n \Gamma(p_n)} \quad (4.52)$$

It can be obtained from the above formula that the peak value of the electron emission coefficient of each part is obtained at $E = (p_n - 1)\varepsilon_n$. Furthermore, if all P_n are equal and ε_n is equal, it can be further simplified to:

$$\frac{d\delta_{ts}}{dE} = \delta_{ts}(E_0) \frac{\left(\dfrac{E}{\varepsilon}\right)^{p-1} e^{-E/\varepsilon}}{\varepsilon \Gamma(p)}, \quad (4.53)$$

where $p = p_n, \varepsilon = \varepsilon_n$. This result indicates that if E_0 is higher and does not depend on n, the function $f_n(E)$ can be determined using the energy spectrum. The peak value of $\dfrac{d\delta_{ts}}{dE}$ is at $E = (p-1)\varepsilon$.

The accumulated energy spectrum can be easily calculated, which can be obtained from equation (4.14) and equation (4.21):

$$S(E_0, E) = \sum_{n=1}^{\infty} n \int_E^{E_0} dE_1 f_n(E_1) \int_0^{\infty} dE_2 \dots dE_n f_n(E_2) \dots f_n(E_n) \theta\left(E_0 - \sum_{k=1}^{n} E_k\right)$$

$$= \sum_{n=1}^{\infty} n \int_E^{E_0} dE_1 f_n(E_1) P_{n-1}(E_0 - E_1)$$

$$= \int_E^{E_0} dE_1 \{f_{1,e}(E_1) + f_{1,r}(E_1)\} + S_{ts}(E_0, E) \quad (4.54)$$

where the TSE part can be expressed as:

$$S_{ts}\left(E_0, E\right) = \sum_{n=1}^{\infty} \frac{nP_{n,ts}\left(E_0\right)}{\Gamma\left(p_n\right)P\left(nP_n, \dfrac{E_0}{\varepsilon_n}\right)} \int_{\frac{E}{\varepsilon_n}}^{\frac{E_0}{\varepsilon_n}} dy\, y^{p_n - 1} e^{-y} P\left[\left(n-1\right)p_n, \frac{E_0}{\varepsilon_n} - y\right] \quad (4.55)$$

Equation (4.54) can be obtained by calculating $\dfrac{\partial}{\partial E}$ on equation (4.52).

4.5.4 SEE Calculation in Multipactor Simulation

In multipactor simulation, the SEE is calculated according to the following steps:

a. Determine the impact energy E_0, position (x, y, z) and angle θ_0 between the incident electrons and normal direction of the material surface.

b. According to the previous derivation, use the following formula to calculate the SEYs of the three types of SEs:

$$\left.\begin{aligned} &\delta_e\left(E_0, \theta_0\right) = \delta_e v \times \left[1 + e_1\left(1 - \cos^{e_2}\theta_0\right)\right] \\[1em] &\delta_e\left(E_0, 0\right) = P_{1,e}\left(\infty\right) + \frac{\left[\hat{P}_{1,e} - P_{1,e}\left(\infty\right)\right]e^{-\left(\frac{\left|E_0 - \hat{E}_e\right|}{w}\right)^p}}{p} \end{aligned}\right\} \quad \text{Ejection} \quad (4.56)$$

$$\left.\begin{aligned} &\delta_r\left(E_0, \theta_0\right) = \delta_r\left(E_0, 0\right) \times \left[1 + r_1\left(1 - \cos^{r_2}\theta_0\right)\right] \\[1em] &\delta_r\left(E_0, 0\right) = P_{1,r}\left(\infty\right)\left(1 - e^{-\left|\frac{E_0}{E_r}\right|^r}\right) \end{aligned}\right\} \quad \text{Scattering} \quad (4.57)$$

$$\left.\begin{aligned} &\delta_{ts}\left(E_0, \theta_0\right) = \hat{\delta}_{ts}\left(\theta_0\right)D\left(E_0 / \hat{E}(\theta_0)\right) \\[1em] &D(x) = \frac{sx}{s - a + x^s} \\[1em] &\hat{\delta}(\theta_0) = \hat{\delta}_{ts} \times \left[1 + t_1\left(1 - \cos^{t_2}\theta_0\right)\right] \\[1em] &\hat{E}(\theta_0) = \hat{E}_{ts} \times \left[1 + t_3\left(1 - \cos^{t_4}\theta_0\right)\right] \end{aligned}\right\} \quad \text{TSE emission} \quad (4.58)$$

Where parameters e_1, e_2, r_1, r_2, t_1, t_2, t_3, t_4 are set according to the material properties.

c. Use formula (4.42) and formula (4.45) to calculate the electron emission probability of n SEs.

d. Using Monte Carlo method, assuming that R is uniformly distributed in $U(0,1)$, calculate in which interval of P_n that R locate in. Then, calculate the number of the SEs n:

When $n = 0$, which means the incident particle is absorbed;

When $n = 1$, there are three possible situations: ejection, scattering and TSE emission. The energy generated is less than E_0, which can be obtained by using the probability density distribution function $f_{1,e}(E) + f_{1,r}(E) + f_{1,ts}(E)$ of energy;

When $n = 2$, only in the case of SEE, the energy satisfies $E_k \in [0, E_0], k = 1, \ldots, n$ and $\sum_{k=1}^{n} E_k \leq E_0$, which can be determined by using the probability density $f_{n,ts}(E)$ of energy.

The angle $0 \leq \phi \leq 2\pi$ between the horizontal direction of the SE and the tangent line coincides with the uniform distribution.

e. Return the velocity and direction of the SE and start the calculation of the next electron.

4.6 SIMULATION OF MULTIPACTOR IN RECTANGULAR WAVEGUIDES

Based on the EM PIC simulation method, the three-dimensional numerical simulation and threshold power analysis of multipactor are conducted for microwave components in various practical applications. More details can be found in references [26–36].

Taking the TE_{10} mode as an example, the long side of the rectangular waveguide cross-section is 20 mm, the short side is 6 mm and the waveguide length is 30 mm. The waveguide wall material is silver, and its maximum SEY is about 2.23. The incident energy corresponding to the maximum SEY is 316 eV, and SEY in the low-energy range (the incident energy is less than about 20 eV) is taken as 0.8. At the initiation stage of simulation, about 20,000 seed electrons (i.e., macro particles, each representing 100 electrons) were loaded and emitted from one of the walls within 1 ns.

For this structure, the simulation frequencies are set as 8 GHz, 10 GHz, 12 GHz and 14 GHz. First, 10 GHz was used as the operating frequency to study the generation, formation, and development of multipactor. When the frequency of input field is 10 GHz and the average power is 140 kW, the phenomenon of resonant electron multiplication occurs. The time evolution of the number of SEs at the initial stage of multipactor is shown in Figure 4.4. SEs are generated from the beginning, and can continue to grow after 4 ns, so the total number of electrons continue to grow (as shown in Figure 4.5).

Furthermore, the current generated by the SEs in the upper and lower walls is analysed. As shown in Figures 4.6 and 4.7, it can be seen that the lower wall first produces relatively high SEE. The reason is that

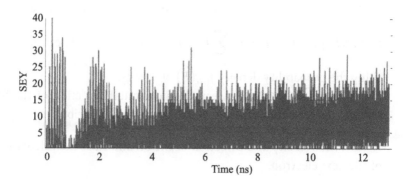

FIGURE 4.4 The change of the total number of SEs with time in the initial stage of multipactor formation.

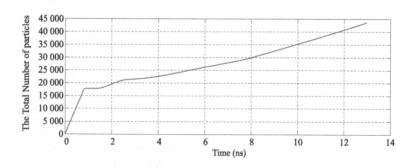

FIGURE 4.5 The change of the total electron number with time in the initial stage of multipactor formation.

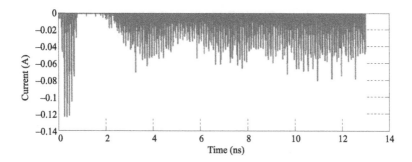

FIGURE 4.6 The current waveform of the SEE from the lower wall in the initial stage of multipactor.

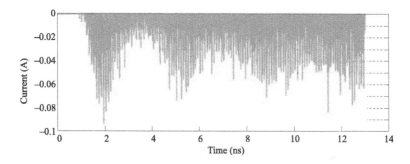

FIGURE 4.7 The current waveform of the SEE from the upper wall in the initial stage of multipactor.

the increase in electric field strength may cause the electrons emit from the surface of the lower wall to return to the lower wall at a higher speed, as shown in Figure 4.8. The kinetic energy of a small number of electrons reaches about 100eV. According to the curve of SEY, when the incident energy is above 100 eV, the SEY is more than 1.5, so the number of SEs will increase significantly. With the development of time, the number of high-energy electrons is increasing, and the number of SEE is also increasing.

Through simulation, the multipactor power threshold (average power) at 4 frequency points of microwave frequency can be obtained, as shown in Table 4.1. The time evolution trend is shown in Figure 4.8. As shown in Table 4.1, in the frequency range from 7.5 to 15 GHz, the multipactor threshold power is increasing.

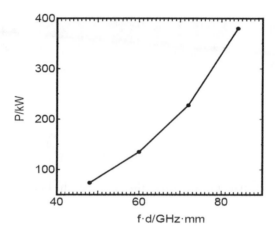

FIGURE 4.8 Threshold power curve of TE_{10} mode in rectangular waveguide.

TABLE 4.1 Threshold Power (Average Power) at Different Frequencies in TE_{10} Mode of Rectangular Waveguide

Frequency/GHz	8	10	12	14
Threshold power /kW	73.5	135	227.5	380

4.7 SIMULATION AND ANALYSIS OF MULTIPACTOR IN AN IMPEDANCE TRANSFORMER

4.7.1 Geometric Modelling

Before the simulation of multipactor in the impedance transformer, the physical structure of the impedance transformer should be established. The outer wall of the waveguide is made of metallic silver. The actual physical dimension of the outermost metal wall at the waveguide port is $26.86 \times 50 \times 8$ mm, and the actual physical dimension of the inner wall at the waveguide port is $22.86 \times 50 \times 5.44$ mm. The inner wall at the waveguide port is the loading area of the input EM field, and the accuracy of its dimension will directly affect the accuracy of the port loading. Therefore, it is necessary to measure the port loading area correctly during the model building process.

4.7.2 Meshing

As shown in Figures 4.9 and 4.10, a uniform meshing method is used in three directions of the rectangular coordinate system. The mesh spacing in three directions is $0.32 \times 0.32 \times 0.32$ mm. In addition, in order to

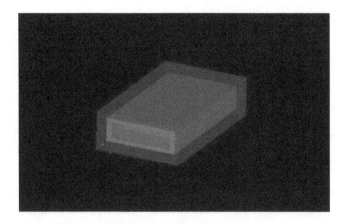

FIGURE 4.9   Meshing of the impedance transformer.

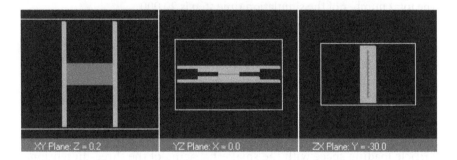

FIGURE 4.10   Sectional view of each coordinate plane.

additionally load a perfectly matched absorption layer (PML) on the wave-guide port, an additional eight air layers need to be built outside the outer wall of the waveguide to place the PML layer.

The meshing results are listed in Table 4.2–4.4:

4.7.3 Simulation Results

The excitation frequency of the multipactor simulation in the above-mentioned waveguide is 9.5 GHz (period of 0.105 ns), and the excitation mode is TE_{10} mode. The material of the waveguide wall is set as lossy silver, and the

TABLE 4.2   Global Meshing Size

Global Dimension Size (X)	Global Dimension Size (Y)	Global Dimension Size (Z)
100	172	41

TABLE 4.3   Meshing Size of Solid Part

Object X_{min}	Object X_{max}	Object Y_{min}	Object Y_{max}	Object Z_{min}	Object Z_{max}
9	93	9	165	9	34

TABLE 4.4   Meshing Size of Air Guide Port Loading Area

WGPAir X_{min}	WGPAir X_{max}	WGPAir Y_{min}	WGPAir Y_{max}	WGPAir Z_{min}	WGPAir Z_{max}
16	86	9	165	14	29

SEE model adopts the modified Vaughan model, the key parameters of the model are $W_m = 165$ eV, $\sigma_m = 2.22$, and $W_0 = 12.5$ eV. The iteration time step is 0.616×10^{-15} s. In each time step, it emits about 80 microparticles or particles in the waveguide transformer. Each particle represents 100 electrons. The total number of the simulation time step is 20,000.

The three-dimensional spatial distribution of the electromagnetic field amplitude evolution with time in the impedance transformer is shown in Figures 4.11 and 4.12. With the evolution of time, the electromagnetic field is evenly transmitted in the impedance transformer and the maximum field amplitude is distributed at the smallest section in the waveguide.

As shown in Figure 4.13, the initial particles are randomly loaded in the particle loading space with the smallest spacing of the three-dimensional model of the impedance transformer. The three-dimensional multipactor simulation is carried out to obtain the image of the particles' evolution in the calculation space. It can be observed that with the expansion of the phase selection effect in the initial stage, the particles gradually aggregate at certain specific phases, and multipactor occurs. Record the number of electrons at each time step under different input powers, and the time evolution trend of the total number of particles is shown in Figures 4.14–4.16. The threshold power of multipactor can be prejudged according to the trend that the total number of electrons increases or decreases exponentially with time. The multipactor threshold power of the impedance transformer is 5600 W.

Electromagnetic calculation results

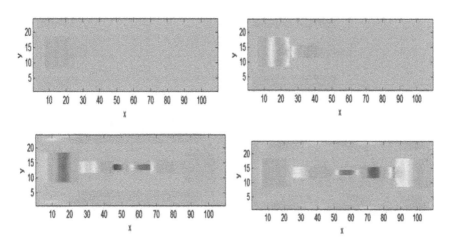

FIGURE 4.11   Electromagnetic field distribution in the XoY plane of the impedance transformer.

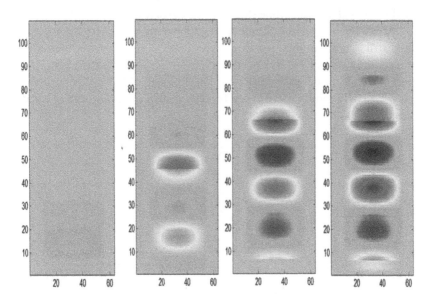

FIGURE 4.12   Electromagnetic field distribution in the XoZ plane of the impedance transformer.

Joint simulation results of electromagnetic particle multipactor

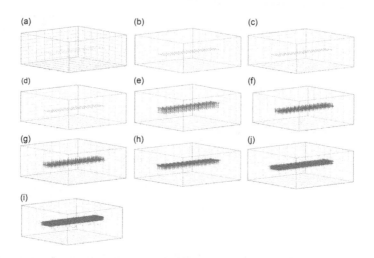

FIGURE 4.13 Three-dimensional distribution of particles in the impedance transformer. (a) Particle distribution with iteration time step of 20, (b) particle distribution with iteration time step of 80, (c) particle distribution with iteration time step of 160, (d) particle distribution with iteration time step of 240, (e) particle distribution with iteration time step of 500, (f) particle distribution with iteration time step of 740, (g) particle distribution with iteration time step of 800, (h) particle distribution with iteration time step of 1100, (i) particle distribution with iteration time step of 1800 and (j) particle distribution with iteration time step of 2800.

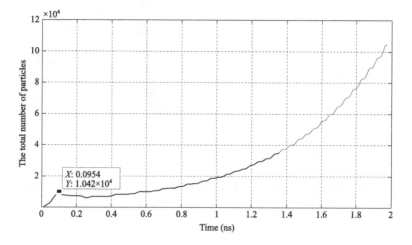

FIGURE 4.14 Multipactor occurs in the impedance transformer under the excitation electric field strength of $2.4\,\text{kV}\,\text{m}^{-1}$.

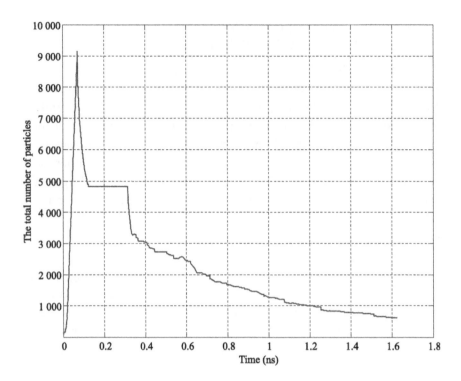

FIGURE 4.15   The time evolution of the number of particles.

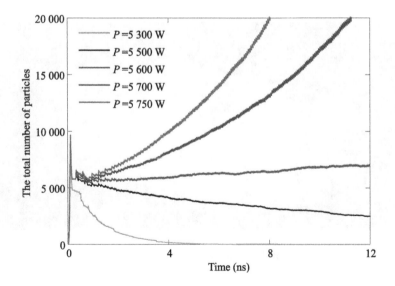

FIGURE 4.16   The time evolution of the number of particles under different excitation power in the impedance transformer.

4.8 SIMULATION AND ANALYSIS OF MULTIPACTOR IN AN RIDGE-WAVEGUIDE FILTER

4.8.1 Geometric Modelling

Before simulating the ridge-waveguide filter, the physical structure should be established. The outer wall of the waveguide is made of metal, and the narrowest section of the ridge-waveguide filter is set as the particle loading region Figure 4.17. The physical size of the outer wall of the waveguide is $70 \times 40 \times 168.1324\,mm$, and the actual physical size of the inner metal wall is $58.17 \times 29.08 \times 168.1324\,mm$. The outer layer is the loading area of the waveguide port, and its size accuracy will directly affect the accuracy of the port.

4.8.2 Meshing

As shown in Figures 4.18–4.21, first, a non-uniform meshing method is used in y direction, the uniform meshing method is used in x and z direction, and the basic spacing in the three directions is $1.28 \times 2.56 \times 0.64\,mm$. In addition, in order to add a PML on the waveguide end, another seven layers of air layers need to be built outside the waveguide outer wall (Table 4.5–4.7).

FIGURE 4.17   Meshing of the ridge-waveguide filter.

FIGURE 4.18   Meshing in the XoY plane.

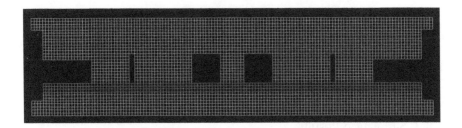

FIGURE 4.19   Meshing in the YoZ plane.

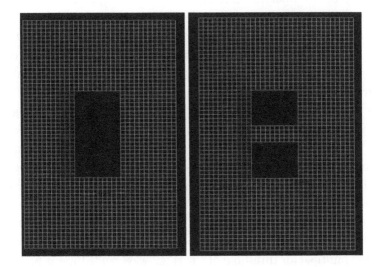

FIGURE 4.20   Meshing in the XoZ plane.

List of meshing results:

TABLE 4.5   Global Meshing Size

Global Meshing Size in x Direction	Global Meshing Size in y Direction	Global Meshing Size in z Direction
71	130	78

TABLE 4.6   Meshing Size of Solid Part

X_{min}	X_{max}	Y_{min}	Y_{max}	Z_{min}	Z_{max}
9	64	9	123	9	71

TABLE 4.7   Meshing Size of Loading Area in Air Waveguide Port

X_{min}	X_{max}	Y_{min}	Y_{max}	Z_{min}	Z_{max}
14	58	9	123	19	62

4.8.3 Simulation Results

For the ridge-waveguide filter, the single-carrier signal excitation frequency of multipactor simulation is 3.9 GHz (period of 0.256 ns), and the waveguide port excitation mode is the TE_{10} mode. The material of the waveguide wall is silver. The time step set in the simulation is 1.23×10^{-15} s. It emits about 7000 particles between two surfaces of the middle ridge. Each particle represents 100 electrons. The total number of the simulation time step is 40,000.

The spatial distribution of normalized intensity of the electromagnetic field in the ridged waveguide filter is obtained by FDTD calculation, as shown in Figures 4.21 and 4.22. The distribution of particles in the YoZ plane of the ridge-waveguide filter is shown in Figure 4.23. It can be seen that with the simulation time increasing, particles gather at the narrowest ridge-waveguide in space. The simulation shows the evolution curve of the number of particles with time under different input power, as shown in Figure 4.24. According to the time evolution trend of the total number of particles, multipactor threshold of the ridge-waveguide filter is prejudged to be between 1300 and 1350 W.

Electromagnetic calculation results

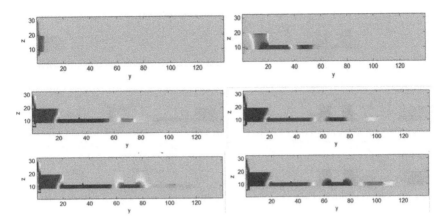

FIGURE 4.21   Electromagnetic field distribution in the YoZ plane of the ridge-waveguide filter.

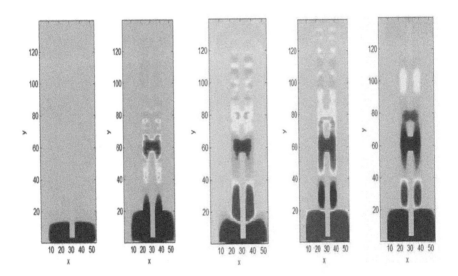

FIGURE 4.22   Electromagnetic field distribution in the XoY plane of the ridge-waveguide filter.

Electromagnetic and particle co-simulation calculation results

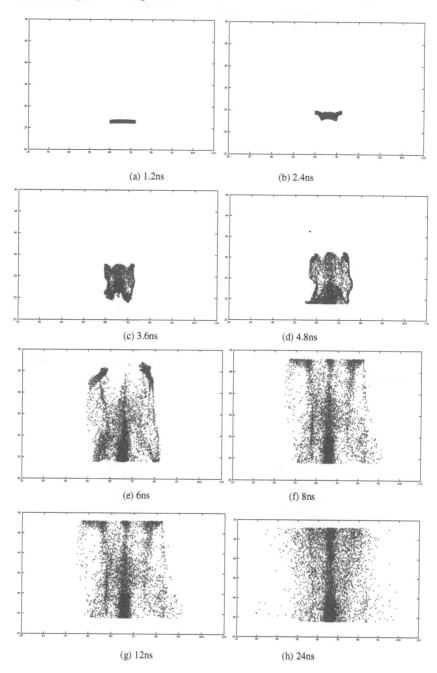

FIGURE 4.23 Distribution of particles in the YoZ plane of the ridge-waveguide filter.

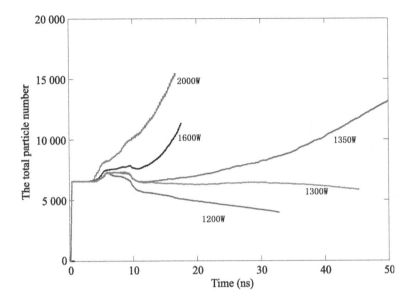

FIGURE 4.24   Particle number/time diagram of the ridge-waveguide filter under different excitation power.

4.9 SIMULATION AND ANALYSIS OF MULTIPACTOR IN MICROWAVE SWITCH

As the basic component in the payload system of the spacecraft, the main function of the microwave switch is to converse the channel state. It realizes more connection states and makes backup for key components in the system. When the key components (such as receivers, high-power amplifiers) fail, it is switched to a backup and improves the reliability of the system operation.

Under vacuum conditions, the resonant electron multipactor effect that occurs in the metal gap of high-power devices has become a key technical bottleneck of microwave switch for high-power space applications.

Microwave switches are usually characterized by the curved surface, minor connection structure, high inhomogeneity of electromagnetic field and so on. In this section, the numerical simulation of the multipactor in microwave switches is carried out through EM PIC co-simulation.

Figure 4.25 shows the three-dimensional model of a microwave switch. The excitation frequency of a single-carrier signal is 1.25 GHz. The dominant EM mode in the switch is the TEM mode, and the metal material is silver. The model is divided into many hexahedral meshes. If the mesh plane, as shown in Figure 4.25, is the XoY plane, the propagation direction

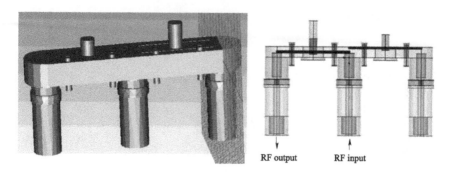

FIGURE 4.25   Three-dimensional model of microwave switch.

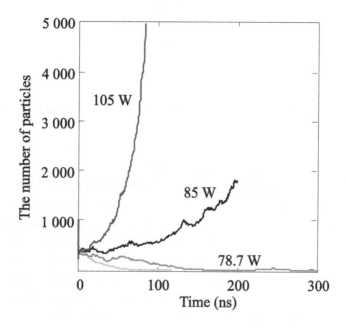

FIGURE 4.26   Curve of particle number with time in microwave switch.

of electromagnetic wave in the switch is along the z direction in the same conversion joint.

In the simulation process, the particles are driven by the electromagnetic force. According to the self-consistent algorithm, the electromagnetic field distribution and particle motion are updated at each time step. When particles collide with the metal wall, the SEE characteristics of the metal material are described using the modified Furman model. The simulation results show the time evolution curves of the particle number under different input power (Figure 4.26). According to the time evolution trend of the total number of particles, multipactor threshold of the switch is prejudged to be between 78.7W and 85W.

In order to supress the SEE of the surface in the microwave switch, which also improves the multipactor threshold power, a porous microstructure has been constructed on the metal surface of the switch. Figure 4.27 shows the scanning electron microscopy (SEM) images of the metal surface under different processing conditions. The SEM image of the metal with a typical machined surface is shown in Figure 4.27a, and the SEM image of the metal with a porous microstructure is shown in Figure 4.27b. The measurement results of the SEY curve on the metal under different surface treatment conditions are shown in Figure 4.28. The typical SEY

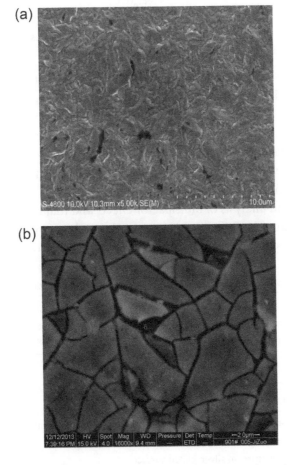

FIGURE 4.27 SEM images of metal samples: (a) with typical machined surfaces and (b) with porous microstructure.

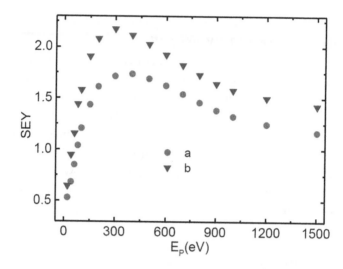

FIGURE 4.28 SEY measurement results of metal samples: (a) with typical machined surface and (b) with porous microstructure.

characteristic parameters that can be used in multipactor simulation are shown in Table 4.8.

Using the EM PIC simulation method, the threshold power of multipactor in the microwave switch with porous microstructure is compared with the microwave switch using the typical machined surface and is shown in Table 4.9. The simulation results show that the electromagnetic and particle joint simulation method is effective and accurate in the numerical simulation and analysis of multipactor effects and that the regular porous microstructures can effectively suppress the multipactor effect of microwave switches and increase the multipactor threshold power.

TABLE 4.8 SEY Parameters of Metal Samples with Different Surface States

Materials	δ_{max0}	E_{max0}	E_1
Metal with typical machined surfaces	2.17	285	29.76
Metal with porous microstructure	1.70	380	35.36

TABLE 4.9 Comparison of Multipactor Threshold Powers in Microwave Switches with Different Surface States

Surface Treatment Method	Simulation Threshold/W
Smooth silver plating	78.7
Porous microstructure on silver-plated surface	166.3

4.10 SUMMARY

In this chapter, based on FDTD and PIC methods of electromagnetic wave introduced in the previous chapters, the SEE model is used to describe the process of interaction between electrons and materials under the drive of electromagnetic wave. The multipactor simulation method is discussed. Multipactor simulation and analysis are carried out for a variety of microwave components, revealing the effectiveness of EM PIC simulation methods in the field of three-dimensional numerical simulation and threshold analysis of multipactor.

REFERENCES

1. Li Y, Cui WZ. Prediction of multipactor thresholds in passive microwave components using an improved simulation method, IEEE Radar Conference, Ottawa, ON, 2013.
2. Li Y, Cui WZ, Zhang N, Wang XB, Wang HG, Li YD, Zhang JF. Three-dimensional simulation method of multipactor in microwave components for high-power space application. *Chinese Physics B*, 2014, 23(4):048402.
3. Taflove A, Hagness SC. *Computational Electrodynamics: The Finite-Difference Time-Domain Method (2nd edition)*. Norwood: Artech House, 2002.
4. Bérenger JP. *Perfectly Matched Layer (PML) for Computational Electromagnetics (1st edition)*. Morgan & Claypool Publishers, San Rafael, CA, 2007:29–35.
5. Hockney RW, Eastwood JW. *Computer Simulation Using Particles*. New York: McGraw-Hill, 1981.
6. Eastwood JW. The virtual particle electromagnetic particle-mesh method. *Computer Physics Communications*, 1991, 64(2):252–266.
7. Goplen B, Ludeking L, Smithe D, Warren G. User configurable MAGIC for electromagnetic PIC calcalations. *Computer Physics Communications*, 1995, 87(1–2):54–86.
8. Liu GZ, Hao S. Space-charge limiting current in spherical cathode diodes. *Chinese Physics B*, 2003, 12(2):204–207.
9. Liu L, Li Y, Wang R, Cui W, Liu C. Particle simulation of low pressure corona discharge in microwave ladder impedance converter. *Chinese Physics*, 2013, 62(2):025201.
10. Kossyi IA, Luk'yanchikov GS, Semenov VE, Zharova NA, Anderson D, Lisak M and Puech J. Experimental and numerical investigation of multipactor discharges in a coaxial waveguide. *Journal of Physics D: Applied Physics*, 2010, 43(34):227–238.
11. Semenov VE, Zharova N, Udiljak R. Multipactor in a coaxial transmission line. II. Particle-in-cell simulations. *Physics of Plasmas*, 2007, 14:033509.
12. Vaughan JRM. Multipactor. *IEEE Transactions on Electron Devices*, 1988, ED-35:1172.

13. Zhang HB, Hu XC, Cao M, Zhang N, Cui WZ. The quantitative effect of thermal treatment on the secondary electron yield from air-exposed silver surface. *Vacuum*, 2014, 102(4):12–15.
14. Chen D. *Communication Satellite Payload Technology*. Beijing: Aerospace Press, 2001.
15. Anza S, Vicente C, Gil J, Mattes M, Wolk D, Wochner U, Boria VE, Gimeno B, Raboso D. Prediction of multipactor breakdown for multicarrier applications: The quasi-stationary method. *IEEE Transactions on Microwave Theory & Techniques*, 2012, 60(7):2093–2105.
16. Marrison AJ, May R, Sanders JD, Dyne AD, Rawlins AD, Petit J. A study of multipaction in multicarrier RF components. Report on AEA/TYKB/31761/01/RP/05 Issue 1, ESA/ESTEC, Noordwijk, The Netherlands, 1997.
17. ESA-ESTEC. Multipaction design and test. ESA Standard ECSS-E-20-01A, ESA, Noordwijk, The Netherlands, 2003.
18. Anza S, Vicente C, Gimeno B, Boria VE, Armendariz J. Long-term multipactor discharge in multicarrier systems. *Physics of Plasmas*, 2007, 14 (8):082112.
19. Vdovicheva NK, Sazontov AG, Semenov VE. Statistical theory of two-sided multipactor. *Radiophysics & Quantum Electronics*, 2004, 47(8):580–596.
20. Anza S, Mattes M, Vicente C, Gil J, Raboso D, Boria VE, Gimeno B. Multipactor theory for multicarrier signals. *Physics of Plasmas*, 2011, 18 (3):032105.
21. Song QQ, Wang XB, Cui WZ, Wang ZY, Shen YC, Ran LX. Multicarrier multipactor analysis based on branching levy walk hypothesis. *Progress in Electromagnetics Research*, 2014, 146(1):117–123.
22. Wang X, Zhang X, Li Y, Cui W, Zhang H, Li Y, Wang H, Zhai Y, Liu C. Particle simulation and analysis of multicarrier microdischarge threshold. *Chinese Physics*, 2017, 66(15):157901.
23. Wang X, Li Y, Cui W, Li Y, Zhang H, Zhang X, Liu C. Global threshold analysis of multicarrier microdischarge based on critical electron density. *Chinese Physics*, 2016, 65(4):047901.
24. Wang XB, Shen JH, Wang JY, Song QQ, Wang ZY, Li Y, Wang R, Hu TC, Xia YF, Sun QF, Yin XS, Cui WZ, Zhang HT, Zhang XN, Liu CL, Li CZ, Ran LX. Monte Carlo analysis of occurrence thresholds of multicarrier multipactors. *IEEE Transactions on Microwave Theory & Techniques*, 2017, 65(8):2734–2748.
25. Wang XB, Qi XK, Chen X, Wei H, Wang BX, Guo LC, Sun QF, Cui WZ, Yin XS, Zhang HT, Zhang XN, Li YD, Liu CL, Zhang B, Ye DX, Ran LX. Generation of coherent multicarrier signals for the measurement of multicarrier multipactor. *IEEE Transactions on Instrumentation & Measurement*, 2017, 66(12):3357–3363.
26. Li Y, Wang D, Yu M, He YN, Cui WZ. Experimental verification of multipactor discharge dynamics between ferrite dielectric and metal. *IEEE Transactions on Electron Devices*, 2018, 65(10):4592–4599.

27. Li Y, Ye M, He YN, Cui WZ, Wang D. Surface effect investigation on multipactor in microwave components using the EM-PIC method. *Physics of Plasmas*, 2017, 24:113505.
28. Li Y, Cui WZ, He YN, Wang XB, Hu TC, Wang D. Enhanced dynamics simulation and threshold analysis of multipaction in the ferrite microwave component. *Physics of Plasmas*, 2017, 24(2):023505.
29. Li Y, Cui WZ, Wang HG. Simulation investigation of multipactor in metal components for space application with an improved secondary emission model. *Physics of Plasmas*, 2015, 22(5):1172–2126.
30. Cui WZ, Li Y, Yang J, Hu TC, Wang XB, Wang R, Zhang N, Zhang HT, He YN. An efficient multipaction suppression method in microwave components for space application. *Chinese Physics B*, 25(6):068401.
31. You JW, Zhang JF, Cui TJ. Power-to-Amplitude Transformation (PAT) for the analysis of guide-wave system based on FDTD method, 2012 IEEE International Workshop on Electromagnetics, Applications and Student Innovation (iWEM), 2012:1–4 .
32. You JW, Wang HG, Zhang JF, Cui WZ, Cui TJ. The conformal TDFIT-PIC method using a new Extraction of Conformal Information (ECI) technique. *IEEE Transactions on Plasma Science*, 2013, 41(11):3099–3108.
33. You JW, Wang HG, Zhang JF, Tan SR, Cui TJ. Accurate numerical method for multipactor analysis in microwave devices. *IEEE Transactions on Electron Devices*, 2014, 61(5):1546–1552.
34. You JW, Wang HG, Zhang JF, Tan SR, Cui TJ. Accurate numerical analysis of nonlinearities caused by multipactor in microwave devices. *IEEE Microwave and Wireless Components Letters*, 2014, 24(11):730–732.
35. You JW, Wang HG, Zhang JF, Li Y, Cui WZ, Cui TJ. Highly efficient and adaptive numerical scheme to analyze multipactor in waveguide devices. *IEEE Transactions on Electron Devices*, 2015, 62(4):1327–1333.
36. Li YD. Electromagnetic PIC numerical simulation technology and its application in high power microwave devices. Doctor thesis. Xi'an Jiaotong University, 2005.

CHAPTER 5

Multipactor Analysis in Multicarrier Systems

5.1 INTRODUCTION

Multipactor is an electron avalanche-like discharge occurring in microwave devices operating at high-power levels and in vacuum or near vacuum condition [1–3]. When initially discovered, it was studied as a beneficial effect for signal amplification in a cold-cathode tube for TV applications by Farnsworth [4], who originally coined the name "multipactor". Today, multipactor is considered as a dangerous collateral effect in high-power vacuum applications, which must be avoided. Unexpected occurrence of multipactors may severely hamper the operation of high-power vacuum devices in space and accelerator applications, leading to interferences to signals, or even hardware damages. The prediction and prevention of multipactor are therefore of vital importance [5, 6].

The phenomenon occurs when initial free electrons (primary) are accelerated by the RF fields and impact against the device walls with enough energy to extract more electrons (secondary) from the surface. If the resulting electronic bunch enters in resonance with the field, this process repeats itself until the electron density reaches a certain level to produce noticeable disturbance of the signal, such as distortion, additive noise or reflection, and ultimately produces a destructive discharge that can even

DOI: 10.1201/9781003189794-5

damage the device. In operation, primary electrons come from different sources such as field emission or electron cascades produced by cosmic rays [7]. In the laboratory, in order to induce the discharge for multipaction testing purposes, different electron seeding techniques are available, such as radioactive sources, controlled electron beams or photoelectric effect [8]. Multipactor may appear in many types of components, such as passive or active high-power devices in guided or microstrip technologies and antennas. Thus, it affects different industry sectors such as satellite communications or particle accelerators.

The biggest effort of the multipactor research lines is devoted to the study and characterization of the phenomenon in order to predict under which conditions it will appear and thus design multipactor-free components. Traditionally, multipactor has been studied for single-carrier signals. The single-carrier prediction techniques are usually based on the multipactor theory, for which there are abundant references, and 2-D or 3-D numerical particle-in-cell (PIC) codes, which combine electromagnetic (EM) solvers and electron trackers. Given some input parameters, such as the frequency of operation, device dimensions and material secondary emission yield (SEY) properties, these single-carrier prediction methods provide the threshold for the multipactor breakdown power. The predicted thresholds are used by the industry to design and assess the margins of operated power in the device to be multipactor free.

Nevertheless, realistic satellite communication systems combine more than one channel in a single output, what is called a multicarrier signal. The multicarrier signal combines the transmission power of the individual channels. Its amplitude is time varying and depends on the relative amplitudes and phases of the channel carriers. Therefore, in the multicarrier path of the spacecraft (after multiplexing the channels), extremely high peak power levels may be attained, thus increasing the risk of a multipactor discharge.

In the past decades, multipactor discharges have been extensively studied. Criteria to evaluate the risk of multipactor discharges were proposed. Examples include the multipactor susceptibility for narrowband single-carrier systems and the N^2 and 20-gap-crossing (20GC) rules for wideband multicarrier systems. Although such criteria may be conservative or lack clear physical understanding, "multipactor-free" devices were engineered following these criteria.

5.2 MULTICARRIER SIGNALS

Modern satellites operate in a multicarrier mode, i.e., several signals at different frequencies exist simultaneously in the microwave and electronic systems. For example, a 15-channel multicarrier signal with the frequency from 11.7GHz to 12.5 GHz, Assuming that each channel has a power of 200 W. Then, the maximum instantaneous power of the system, the peak power, is equal to 45 kW. The peak power increases with the square of the number of carriers. Such a high instantaneous power is very unlikely in a real system since it will occur only when all the signals are in phase, as illustrated in Figure 5.1. The most likely scenario, if the carriers are not phase locked, is that the phase of each carrier is a random number and will result in a signal with much lower maximum instantaneous power, as illustrated in Figure 5.2 [9].

An arbitrary multicarrier signal $V(t)$ can be described as:

$$V(t) = \sum_{i=1}^{N} A_i \cos\left(2\pi\left(f_1 + \sum_{j=1}^{i-1} \Delta f_j\right)t + \varphi_i\right), \tag{5.1}$$

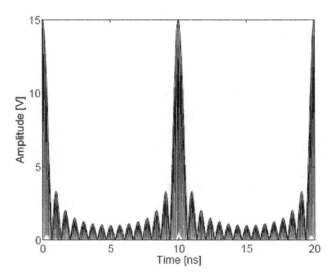

FIGURE 5.1   In-phase multicarrier signal. The signal oscillates rapidly, which makes the signal envelope appear clearly in this time resolution.

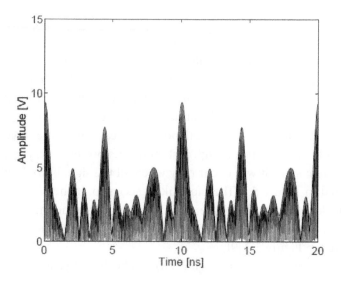

FIGURE 5.2   Random phase multicarrier signal.

where N is the number of carriers, A_i and φ_i are the amplitude and initial phase of the i-th carrier, f_1 is the frequency of the first carrier and Δf_j is the frequency spacing between the i-th and $(i+1)$-th carriers. For convenience, we denote the initial phase combination of this N-carrier signal as $\boldsymbol{\Phi}_i = [\varphi_{i1}, \varphi_{i2}, \ldots, \varphi_{iN}]$. Note that the envelope of $V(t)$ is periodic, whose period is the least common multiple of the periods of all carriers. For example, Figure 5.3 plots the waveforms within a single envelope period of two typical multicarrier signals with "in-phase" and "triangular-phase" combinations of φ_i. It is seen that with the same A_i, f_1 and Δf_j, different $\boldsymbol{\Phi}_i$ could result in waveforms with very different peak-to-average ratios (PARs), each with different "single-event" multipactor discharge (SMD) and "long-term" multipactor discharge (LMD) thresholds.

The signals in Figures 5.1–5.3 are characterized by all carriers having the same amplitude and a constant frequency. Consider a signal with N carriers, with each carrier having the same amplitude E_0 but different phases ϕ_n and with a frequency spacing Δf. The period of the envelope will then be $T = 1/\Delta f$, and the envelope is given by

$$E_{\text{env}} = E_0 \sqrt{\left(\sum_{n=0}^{N-1} \cos\left(n2\pi\Delta ft + \phi_n \right) \right)^2 + \left(\sum_{n=0}^{N-1} \sin\left(n2\pi\Delta ft + \phi_n \right) \right)^2} \quad (5.2)$$

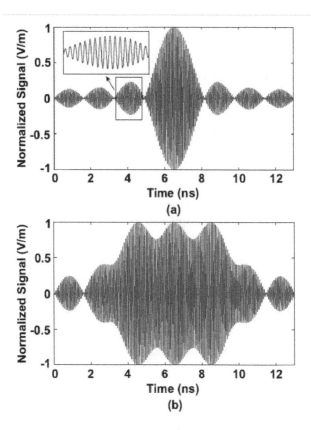

FIGURE 5.3 Waveforms of an in-phase (a) and a triangular-phase (b) multicarrier signal simulated with $A_i=1$, $f_1=11.744$ GHz and $\Delta f=0.0766$ GHz.

A more realistic signal would have different amplitudes for each carrier, and the frequency spacing would not be constant. The envelope of such a signal can be found from

$$E_{\text{env}} = \sqrt{\left(\sum_{n=0}^{N-1} E_n \cos\left(k_n \omega_0 t + \phi_n\right)\right)^2 + \left(\sum_{n=0}^{N-1} \sin\left(k_n \omega_0 t + \phi_n\right)\right)^2} \qquad (5.3)$$

where k_n is a factor determining the frequency spacing

$$k_n = \frac{f_n}{f_0} - 1, \; n = 0,1,\ldots,N-1 \qquad (5.4)$$

f_0 is the lowest carrier frequency and $\omega_0 = 2\pi f_0$. When assessing the worst case scenario from the multipactor point of view, it is important to study a whole envelope period. For arbitrarily spaced frequencies, the envelope period, T, can be found by solving the following Diophantine systems of equations:

$$T = \frac{n_i}{\Delta f_i}, n_i \in N, \ i = 1,2,\ldots,N-1 \tag{5.5}$$

where N is the number of carriers, f_0 is the signal with the lowest frequency and $\Delta f_i = f_i - f_0$. The envelope period will be the solution with the smallest possible integers. For equally spaced carriers, the solution becomes $n_1 = 1$, $n_2 = 2, \ldots, n_{N-1} = N-1$, which implies that $T = 1/\Delta f$, like before.

When studying the multicarrier multipactor, it is common to make certain simplifications that will allow using a single-carrier methodology to asses also the multicarrier case, e.g., the mean frequency of all the carriers is used as the design frequency. Thus, most of what has been said about single-carrier multipactor will then be valid also for the multiple signal cases.

From an industrial point of view, it is important not only to understand the physics of multipactor but also how the theoretical and experimental results should be applied when making multipactor-free microwave hardware designs. In Europe, most space hardware designers follow the standard issued by ESA [8]. This standard includes both the single and the multicarrier cases, but for the latter, it is stated that the design guidelines are only recommendations. Most research support these recommendations, but not enough tests have been performed to verify the theoretical findings. When using the standard, it is important to be aware of the fact that it is primarily based on the parallel-plate model with a uniform electric field. Design with respect to this approach for other geometries is normally a conservative and safe way. However, in many common microwave structures, the geometry is such that losses of electrons are much higher than that in the parallel-plate case. Thus, the multipactor threshold in geometries such as coaxial lines, waveguides and irises can be higher or much higher than that obtained using the plane-parallel model.

5.3 TWENTY GAP-CROSSING RULE (TGR)

In the multicarrier case, only components of type 1 are covered by the recommendations given in the ESA standard [8]. Type 2 and type 3 components will require further research before they can be included in the standard. In the single-carrier case, the level that is compared with the multipactor threshold in the susceptibility chart is the amplitude of the signal and no ambiguities exist. For multicarrier designs, the traditional way of designing was to set the design margin with respect to the peak power of in-phase carriers, as shown in Figure 5.1. This design method is still allowed by the ESA standard, and for type 1 components, the design margins range from 0 to 6 dB depending on the type of testing that will be performed. However, as previously mentioned, in-phase carriers for non-phase locked signals are extremely unlikely and thus the standard allows for another design margin, which is set with respect to the so called P20 power level. The P20 level corresponds to the "peak power of the multicarrier waveform whose width at the single-carrier multipaction threshold is equal to the time taken for the electrons to cross the multipacting region 20 times" [8]. This level is illustrated in Figure 5.4.

In the case when a design is made with respect to the P20 level, the design margins range from 4 to 6 dB depending on the type of testing. A problem with the P20 level is that it is not a trivial problem to find the peak power level for 20 electron gap crossings. This power level is usually referred to as the worst case scenario, even though it may not always be the worst case from a multipactor point of view. A number of different ways of finding the worst case scenario have been proposed, e.g., using parabolic or triangular phase distribution in the equally spaced carrier case. Some of the better methods for finding the worst case scenario will be described in the following subsections after a brief discussion.

The TGR was proposed in Ref. [10] in 1997 and in its original version it reads:

> As long as the duration of the multicarrier peak and the mode order of the gap are such that no more than twenty gap-crossings can occur during the multicarrier peak, then multipaction-generated noise should remain well below thermal noise (in a 30 MHz band).

The rule is a result of an analysis of simulated multicarrier multipactor discharges. Comparison with experiments showed great deviations, where the simulated noise could be as much as 75 dB greater than the measured noise level. In the experiments, a minimum of 99 gap-crossings were required before the produced noise was detectable above the noise floor of −70 dBm. Of course, there may be bit errors even at lower noise levels, but as the number of electrons grows exponentially with the number of gap crossings, there is a huge difference between 20 and 99 gap-crossings.

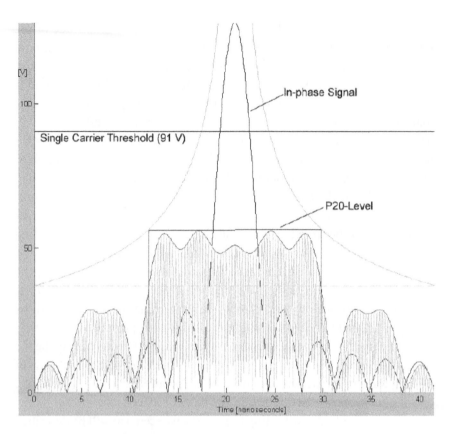

FIGURE 5.4 An example where the in-phase peak power is above the single-carrier threshold, while the P20 level is more than 4 dB below the same threshold. The peak voltage is 128.4 V, the single-carrier threshold is 91 V, and the P20 voltage is 57 V. Signal data: 12 carriers, equally spaced, $f_{min} = 1.545$ GHz, $\Delta f = 24$ MHz and each carrier amplitude is 10.7 V. Material properties: $W_1 = 23$ eV and $\sigma_{se, max} = 3$.

However, the TGR is certainly a good first attempt to lower the requirements for multicarrier multipactor. It is a fairly conservative method and thus the risk of applying it should be quite limited. However, more appropriate guidelines should be based on an unambiguous theoretical concept, which can take the material properties into account. Then, when performing simulations and experiments to verify the idea, it is of paramount importance to make sure that the actual material properties of the test samples are well known and that these properties are also being used in the simulations. Due to the large difference in secondary emission properties between different materials, it would seem reasonable that for a material with a low $\sigma_{se,\,max}$ one would allow more gap crossings than, in the opposite case, for a material with a high $\sigma_{se,\,max}$.

5.4 LONG-TERM MULTICARRIER MULTIPACTOR

A new mechanism of multipactor discharge is found in multicarrier systems [11]. In the process of a multicarrier multipactor discharge (MMD), the fluctuating space electrons (SEs) between consecutive envelope periods can be accumulated, yielding a LMD . Compared with the SMD induced by the sustained high-power envelope inside a single envelope period, an LMD can be induced by lower-energy fields, and thus conventional criteria no longer apply for LMDs. The potential occurrence of LMDs increases the risk of multipactors in wideband systems and hence has attracted much attention .

It is straightforward that the LMD threshold can be obtained by investigating the fluctuation of SEs between successive envelope periods of a multicarrier signal. Achieving an efficient approach to rapidly compute such fluctuations is therefore essential in both SMD and LMD studies. In order to depict this fluctuation inside and between successive envelope periods, researches based on theoretical analysis and PIC simulations have been conducted.

Among many researches on LMDs, representative theoretical analyses are reported by S. Anza et al. [12]. By removing the assumption of stationary state and introducing probability analysis based on transit time probability density, the fluctuation of SEs in homogeneous fields can be theoretically computed [13]. This theory has been extended to analyse the fluctuation of SEs in inhomogeneous fields by deriving the probability density of the lateral diffusion of SEs based on branching Levy walk theory [14]. However, currently these theoretical approaches may not be applicable for practical, inhomogeneous microwave components.

In comparison, PIC simulations are able to analyse arbitrary devices. Commercial tools such as FEST3D™, SPARK3D™ and CST Particle Studio™ have been widely used in simulating MMDs [15–17]. However, such simulations suffer from inconceivable time consumption and extreme requirement on computational capacity. It is inapplicable for stochastic optimization-based threshold analysis of MMDs.

In [18], the ideas of simplification of multicarrier signal, fast evaluation of electron growth and the use of global optimizer to find the global threshold of LMDs were proposed based on an quasi-stationary (QS) analysis. In [19], an efficient approach to rapidly compute the fluctuation of space electrons in the evolving process of a multicarrier multipactor is proposed. Based on this, Monte Carlo approaches are developed to efficiently find the global "worst case" waveforms and the occurrence thresholds for both "single event" and "long-term" multipactors. In particular, an "x-gap-crossing" (XGC) method is proposed for the investigation of SMDs. Based on the concept of the "detectable threshold" [8], the connection between the "worst case" waveform and the number of SEs possibly accumulated inside a single envelope period is established. By doing so, the threshold power, the "worst case" waveform and the actual "On interval" of an SMD can be obtained. It shows that not only the transition time of the "20-gap-crossing" (T20) is not special compared with the other time windows, but also the 20GCR is also very conservative compared with the optimized "On interval". It reveals that there exists a "critical bandwidth" for a multicarrier signal, which determines the type (i.e., SMD or LMD) of an occurred discharge. It can also be used to find the "best case" waveform of a signal that has the highest threshold power of an MMD, implying a prospect to significantly decrease the risk of MMDs. The proposed approaches can be used for wideband devices, providing an efficient, integrated solution for the analysis of MMD for practical devices with inhomogeneous field distribution.

Before the discovery of the LMD, the empirical 20GCr is normally used to estimate the MMD thresholds. It says that if any continuous segment of the envelope of $V(t)$ is higher than the discharge threshold of a single-carrier signal with the lowest frequency among all the carriers, during which SEs are able to transit 20 times between the device's surfaces, an MMD could potentially occur . Regardless of the inaccuracy concern about this rule, it only considers the accumulation of SEs in a single envelope period and cannot be used to predict LMD thresholds. Thus, the global MMD threshold, which is the smaller one in the SMD and LMD thresholds, cannot be obtained by the 20GCR either.

Since the waveform of a multicarrier signal is determined by Φ_i, to obtain an LMD threshold, the "worst case" waveform that results in the "critical state" of the fluctuation of SEs has to be found. For a given Φ_i, the "critical state" occurs when SEs fluctuate in a way that neither increases nor decreases with time at a specific power of the multicarrier signal. This causes two technical challenges. First, since each φ_i takes values between 0° and 360°, the combinations of φ_i can be prohibitively large. For instance, with a one-degree phase resolution, there could be 4×10^{30} combinations of a 12-carrier signal. Second, existing methods are not efficient enough to calculate such a fluctuation during an MMD. As a result, finding the "worst case" waveform by examining this fluctuation for each Φ_i would cost excessive time.

By the time of speaking, the theory for multipactor and multicarrier signals is rather scarce. To the authors' knowledge, the only existing full theory for multicarrier operation is provided in [12,18,19]. Numerical solvers capable of handling multicarrier signals exist as well. However, in the multicarrier case, there are many more parameters involved in the multipactor discharge than that for the single-carrier case, which include the carrier frequency spacing, the relative phases among the carriers and the amplitude (or power) per carrier. Therefore, the current multipactor theory and numerical software tools for multicarrier signals are able to determine whether there is multipactor discharge for a fixed configuration. However, they do not provide the worst case, which is the combination of all the variables of the problem that produces a multipactor discharge with the minimum power per carrier. Thus, the current multicarrier theory/software tools do not predict the lowest multipactor breakdown level.

The design rules that are currently being applied by the space industry are based on simplifications that allow applying the single-carrier predictions to the multicarrier case. The most restrictive one is K^2, K being the number of carriers, which takes the peak power of the multicarrier signal as the continuous wave (CW) power of an equivalent single-carrier signal. The multipactor breakdown is then equal to P_{sc}/K^2 per carrier, where P_{sc} is the single-carrier breakdown that can be calculated with single-carrier predictors. This design rule is known to be very conservative and typically gives much lower breakdown power predictions than measured ones. This imposes unnecessary constraints on the design and usually forces to carry out cumbersome test campaigns to validate the components.

The first attempt for trying to reduce the margins is the 20-gap-crossing rule (20GCR), which establishes a more relaxed criterion of multipactor. It can only appear when the multicarrier signal envelope is above the single-carrier threshold for a time such that an electron crosses the gap 20 times. In other words, the 20GCR allows the multicarrier signal to be above the threshold for a short time, assuming that the electron build-up will not be enough to produce a discharge. Equivalently, the above K^2 rule would be the zero-gap-crossing rule, i.e., it does not allow any electron crossing (impact) above the threshold. With respect to the rule, the 20GCR predicts higher multipactor thresholds and reduces the design constraints.

However, the 20GCR is based only on the study of numerical simulations and measurements and does not have a solid physical basis. The question that naturally arises is why 20 and not another value, and why 20 should be a universal value valid for all kind of signals and devices. This uncertainty on the prediction rule implies large safety margins that are imposed to the predicted values. As a consequence, the 20GCR, although being more relaxed than the K^2 rule, still yields very conservative predictions in most cases.

Anza proposes a novel QS prediction technique for multipactor in multicarrier signal [18], with the aim of giving more accurate predictions in order to reduce the safety margins, avoid unnecessary design constraints and reduce the test campaigns as much as possible. Even if a full multicarrier theory is already available, the new technique presented in this paper is still based on the single-carrier theory, following a similar approach as the previous ones. However, it takes more sophisticated simplifications on the multicarrier signal and employs a new electron growth model. By applying the single-carrier theory, the number of parameters of the problem reduces significantly and allows for more simple and intuitive solutions.

The QS prediction method is based on the nonstationary theory for single-carrier signals, which belongs to the family of statistical theories that introduce the randomness of electron emission velocity and angle. In spite of their complexity, the statistical theories have the advantage of matching the experimental results better. Among the statistical theories, the nonstationary one is able to model both the electron growth and absorption, above and below the multipactor threshold, and it considers both single- and double-surface interactions. In addition, it gives analytical expressions for the instantaneous SEY and multipactor order. Therefore, the nonstationary theory becomes the most suitable one for multipactor prediction with multicarrier signals.

For a multicarrier signal with a specific set of frequencies, its envelope will have a fixed period, but its shape will vary in accordance with the choice of the phase and amplitude of each carrier. The shape may be seen as a set of periodic lobes. In general, the height of such lobes is related to its width in such a way that the higher the envelope is, the narrower the lobes are. Theoretically, for equal amplitude, the multicarrier signal envelope is between two limit values, KV_0 corresponding to the in-phase scheme (all carriers have the same relative phase), and $\mathrm{sqr}(K)V_0$ for a totally uncorrelated phase scheme (where the lobes spread and overlap to form a flatter envelope). There are different boundary models that relate the height and the width of the envelope, such as Wolk et al. [20] or Angevain et al. [21] boundary functions. These provide the voltage factor F_v, which relates the boundary level and the level per carrier, for each envelope width, ΔT

$$V_e(\Delta T) = F_v(\Delta T)V_0 \qquad (5.6)$$

Figure 5.5 shows an example for different phase schemes for a signal with $K = 8$ and $V_0 = 1$ with a uniform frequency spacing 40 MHz.

The instantaneous frequency is also periodic with the same period of the envelope with an oscillating value around the mean frequency of all carriers. Therefore, the frequency of the multicarrier signal can be approximated as a constant value equal to the mean frequency of all carriers, i.e., f_m.

The study of the multipactor phenomenon in multicarrier signals is rather more complicated than that for the single-carrier case. Conceptually, the process can be described as follows. When the multicarrier signal envelope, $V_e(t)$, surpasses a certain level, the electrons are accelerated with

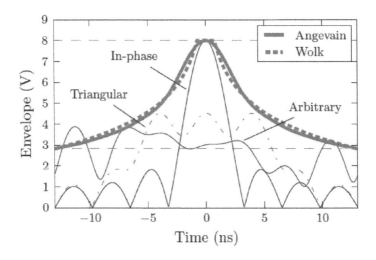

FIGURE 5.5 Multicarrier signal envelopes for different phase schemes with $K = 8$.

enough energy to initiate a multipactor discharge, and thus, the electron population increases. The value of such a threshold is not well known. However, there is evidence that indicates that it must be close to the breakdown threshold in the single-carrier case, V_{sc}, for a frequency equal to the mean frequency of all carriers, f_m. On the other hand, when $V_e(t)$ is below V_{sc}, the electrons impact on the device walls with low energies, implying a SEY below 1, and the electrons being therefore absorbed.

The intervals in which $V_e(t)$ is above V_{sc} are called "on" intervals, and those where it is below are known as "off" intervals [11]. Since the envelope is periodic, "on" and "off" intervals are alternated indefinitely in time. Hence, there will be a multipactor discharge in two cases. Either the "on" interval is long enough to make the electron population grow to a detectable level in the first period of the envelope, which is called a single-event discharge, or the electron growth during the "on" interval is higher than the electron absorption during the "off" interval. This makes the electron population grow slowly, period after period, culminating in a LMD. Figure 5.6 shows an example of a long-term multipactor discharge with an in-phase multicarrier signal, extracted from [12]. The long-term discharge build-up is typically in the range of few nanoseconds and the multipactor discharges are, in general, not self-sustained. Therefore, in practice, both kinds of discharges are indistinguishable in the laboratory. Nevertheless, each of them has different implications for the multipactor breakdown level. Long-term discharges are thought to be more restrictive than single-event ones [18].

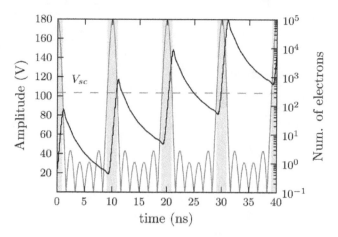

FIGURE 5.6   Example of electron growth in a long-term multipactor discharge extracted from [12].

There are infinite combinations of amplitudes and phases that lead to a multipactor discharge. Assuming that all carriers have equal amplitude, the worst case is defined as *the combination of phases that causes a multipactor discharge with the minimum amplitude (or power) per carrier.* This worst case must be the goal of any multipactor prediction method for multicarrier signals.

The 20GCR is very simple. It simplifies the multicarrier envelope as a pulsed signal, which can only be above ("on") or below ("off") the single-carrier threshold. As its own name indicates, it establishes that there will be a multipactor discharge when the "on" interval is long enough to ensure at least 20 electron impacts. In order to provide a larger margin, the 20GCR takes the lowest frequency of the train of carriers, f_i (instead of f_m), as the reference frequency for the calculation of the single-carrier breakdown threshold. For a multipactor discharge of order n (the order of the multipactor discharge sets the number of cycles between consecutive impacts for a single electron), the "on" time is

$$T_{20} = \frac{10*n}{f_1} \tag{5.7}$$

The rule does not give any value for the worst case phases or RF breakdown power. It just gives the length of the "on" interval. In order to find such a combination of phases and power, it is necessary to conform the envelope to the desired shape through numerical optimizers, such as simulated annealing or genetic algorithms, which search the right combination of phase and amplitude for each carrier ensuring T_{20}. Another possibility is to use boundary functions for the envelope amplitude such as, which only estimates the breakdown power.

The main advantage of this rule, i.e., its simplicity, is at the same time its main drawback. It is an empirical rule, and it is not clear why the criterion that leads to a number of 20GC is applicable to all situations. This is, it neither takes into account how high the envelope with respect to the single-carrier threshold is, nor the dependence of the multipactor order with voltage, or the kind of material in terms of the SEY curve.

For amplitudes close to the breakdown level, the higher the envelope amplitude, the higher the impact energy is, and thus, the higher the SEY. Therefore, it seems logical that for higher amplitudes, the number of necessary impacts to cause a detectable discharge gets lower. For instance, for an amplitude equal to the single-carrier threshold, the SEY is nearly 1, which implies no electron growth (and no discharge) at all, no matter how many electron impacts occur.

On the other hand, it also seems logical that the number of gap crossings to create a discharge is different for materials with different SEY curves. For example, it would be expected that the number of gap crossings for gold would be higher than that for aluminium since gold is known to typically have a much lower SEY. Furthermore, the 20GCR only takes into account single-event discharges and completely disregards long-term discharges.

REFERENCES

1. Gill EWB, von Engel A. Starting potentials of high-frequency gas discharges at low pressure. *Proceedings of the Royal Society A: Mathematical, Physical and Engineering Sciences*, Feb. 1948, 192(1030):446–463.
2. Hatch A, Williams, H. The secondary electron resonance mechanism of low-pressure high-frequency gas breakdown. *Journal of Applied Physics*, Apr. 1954, 25(4):417–423.
3. Vaughan J. Multipactor. *IEEE Transactions on Electron Devices*, Jul. 1988, 35(7):1172–1180.
4. Farnsworth P. Television by electron image scanning. *Journal of The Franklin Institute*, Oct. 1934, 218(4):411–444.
5. Cameron RJ, Mansour R, Kudsia CM. *Microwave Filters for Communication Systems: Fundamentals*, Design and Applications. New York: Wiley, 2007.
6. Geng R, Goudket P, Carter R, Belomestnykh S, Padamsee H, Dykes D. Dynamical aspects of multipacting induced discharge in a rectangular waveguide. *Nuclear Instruments and Methods in Physics Research Section A: Accelerators, Spectrometers, Detectors and Associated Equipment*, 2005, 538(1–3):189–205.
7. Sternglass EJ. Theory of secondary electron emission by high-speed ions. *Physical Review*, Oct. 1957, 108(1):1–12.
8. ESA. *Space Engineering: Multipacting Design and Test*. Noordwijk, The Netherlands: ESA, May 2003, vol. ECSS-20-01A.
9. Udiljak R. *Multipactor in Low Pressure Gas and in Nonuniform RF Field Structures*. Sweden: Chalmers University of Technology, 2007.
10. Marrison, AJ, May, R, Sanders, JD, Dyne, AD, Rawlins, AD, Petit J. A study of multipaction in multicarrier RF components. Noordwijk: ESA/ESTEC, Report on AEA/TYKB/31761/01/RP/05 Issue 1, Jan. 1997.
11. Anza S, Vicente C, Gimeno B, Boria VE, Armendariz J. Long-term multipactor discharge in multicarrier systems. *Physics of Plasmas*, Aug. 2007, 14(8):082112–8.

12. Anza S, Mattes M, Vicente C, Gil J, Raboso D, Boria VE, Gimeno B. Multipactor theory for multicarrier signals. *Physics of Plasmas*, 2011, 18(3):032105.
13. S. Anza, C. Vicente, J. Gil, V. E. Boria, B. Gimeno, and D. Raboso, "Nonstationary statistical theory for multipactor," *Physics of Plasmas*, Jun. 2010, 17(6):062110–11.
14. Song QQ, Wang XB, Cui WZ, Wang ZY, Shen YC, Ran LX. Multicarrier multipactor analysis based on branching levy walk hypothesis. *Progress Electromagnetics Research*, May 2014, 146:117–123.
15. FEST3D full-wave electromagnetic simulation tool. Univ. Politec. Valencia, Valencia, Spain, 2012. [Online]. Available: http://www.fest3d.com.
16. SPARK3D multi-format high power simulation tool. Univ. Politec. Valencia, Valencia, Spain, 2012. [Online]. Available: http://www.fest3d.com.
17. CST. Computer Simulation Technology (CST) Center, Framingham, MA, 2012. [Online]. Available: http://www.cst.de.
18. Anza S, Vicente C, Gil J, Mattes M, Wolk D, Wochner U, Boria VE, Gimeno B, Raboso D. Prediction of multipactor breakdown for multicarrier applications: the quasi-stationary method. *IEEE Transactions on Microwave Theory and Techniques*, Jul. 2012, 60(7):2093–2105.
19. Wang XB, Shen JH, Wang JY, et al. Monte carlo analysis of occurrence thresholds of multicarrier multipactors. *IEEE Transactions on Microwave Theory & Techniques*, 2017, 65(8):2734–2748.
20. Wolk D, Schmitt D, Schlipf T. A novel approach for calculating the multipaction threshold in multicarrier operation. Proceedings of 3rd International ESTEC Multipactor, RF, DC Corona, Passive Intermodulation in Space RF Hardware Workshop, Noordwijk, Sep. 4–6, 2000, 85–91.
21. Angevain J-C, Drioli L, Delgado P, Mangenot C. A boundary function for multicarrier multipaction analysis, 3rd European Conference on Antennas and Propagation, Mar. 2009, 2158–2161.

CHAPTER 6

PIC Simulation of Collector for TWT

6.1 INTRODUCTION

Vacuum electronic devices have been developed since more than 100 years. They are not only used in radio and television stations as an important part of signal transmission sources but also used in radar, communications, electronic countermeasures, telemetry and remote control and precision guidance equipment as the heart of information weaponry [1]. Meanwhile, it is also an important component of the transponder in the communication of spacecraft systems. As one of the key vacuum electronic devices [2], traveling wave tube (TWT) has a wide operating frequency bandwidth, high gain, high power, low noise, and good nonlinear characteristics compared with other types of power amplifiers. In addition, TWT has good performance of reliability. The lifetime of the space traveling wave tube (STWT) [3] can reach 10–15 years even in harsh environments.

A TWT is a type of device that amplifies the high-frequency field by obtaining the energy of electron beam with the interaction between electron beam and high-frequency field. In a TWT, the electron beam must constantly transfer its own kinetic energy to the high-frequency field, and the synchronization conditions that the electron beam and traveling wave field interact must be satisfied. Thus, the efficiency of the electron beam is relatively low in the TWT. The function of the collector is to collect the electron beams which transfer energy to the high-frequency field in the interaction zone. In spacecraft systems, the efficiency of parameter becomes particularly important for STWTs, due to the limited energy of the spacecraft which operates in orbit. Even it can be said that efficiency is the most important parameter for a STWT. As one of the important part of the STWTs, the efficiency of the collector can affect the efficiency of the entire STWT. Therefore, how to improve the efficiency of the collector is also one of the key issues that need to be considered during the development of the STWT. Based on this background, the electromagnetic particle simulation algorithm for the collector of the TWT is introduced in this chapter, and the method to calculate efficiency of collector is also discussed.

6.2 PRINCIPLE OF TRAVELING WAVE TUBE

A TWT mainly includes five parts, such as an electron gun, a slow-wave system, a magnetic focusing system, a collector, and the input/output couplers (Figure 6.1). Their specific functions are described as follows [2].

1. Electron gun. The function of the electron gun is to generate an electron beam with a certain current intensity, and the electron beam is focused into a certain shape. Then, the electron beam is injected into the slow-wave circuit area after the speed is accelerated. Here, the speed of the electron beam is slightly greater than the phase velocity of the traveling wave transmitted in the slow-wave circuit, so that the electron beam and the traveling wave can exchange energy.

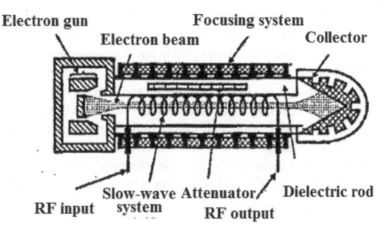

FIGURE 6.1 The schematic diagram of traveling wave tube.

2. Slow-wave system. The slow-wave system is a place where electron beam and high-frequency field interact, and so as to amplify microwave energy. Its task is to transmit high-frequency electromagnetic traveling waves and to reduce the phase velocity of the electromagnetic waves to a speed slightly smaller than that of the electron beam, thereby achieving synchronization conditions. In addition, the slow-wave circuit must provide a strong axial electric field to ensure the intensity of the beam-wave interaction. In order to avoid self-excited oscillation, one or more concentrated attenuators are usually set in the slow-wave circuit. Sometimes, to better isolate the microwave signal in the input section and the output section, the cut-off model is also used in the slow-wave circuit.

3. Magnetic focusing system. The function of the focusing system is to balance the space charge repulsive force inside the electron beam with an external magnetic field force, so as to constrain the electron beam so that it will not diverge to ensure that it can pass through the entire slow-wave circuit smoothly. The currently used focusing system is a periodic permanent magnet focusing system.

4. Collector. The role of the collector is to collect the electrons that have been exchanged with the electromagnetic field. Because the electron beam still has a high speed at this time, part of the energy of the electron beam can be recovered under the action of the depressurizing collector. The energy generated by the electrons hitting the collector is dissipated by thermal energy consumption. Therefore, in order to improve the efficiency of the TWT, a multistage depressed collector is generally used. The inner and outer surfaces of the collector must be specially designed to reduce secondary electron emission and enhance heat dissipation capabilities.

5. Input and output couplers. The function of the input coupler is to feed the high-frequency signal to be amplified to the slow-wave circuit in the TWT, while the function of the output coupler is to couple the high-frequency signal that has been amplified from the slow-wave circuit to the outside of the tube. In general, coaxial structures are used in low-frequency, wideband and low-power areas, while waveguide structures are used in high-frequency and high-power areas.

The operating principle of a TWT is described in this section. First, the electron beam emitted from the cathode of electron gun is injected into the high-frequency structure and shaped by the focusing system. The electron beam which has entered the high-frequency structure interacts with the electromagnetic wave which is fed from the high-frequency input port and transmitted on the high-frequency structure. Then, the electromagnetic wave in the high-frequency circuit is amplified by absorbing the energy from electron beam and is output from the output port. Finally, the electron beam which has lost energy is received by the collector, and the energy remained is dissipated on the collector in the form of thermal energy.

6.3 OPERATING PRINCIPLE OF THE COLLECTOR OF TRAVELING WAVE TUBE

In the TWT, since the electron beam and the traveling wave field can interact effectively only under the condition of synchronization, the energy that the electron beam can deliver to the high-frequency field is only the kinetic energy that is the higher velocity part. Thus, the electronic efficiency is low and only about 20%. In order to improve the efficiency of TWTs, a series of measures have been adopted. Now, there are mainly two types of methods. One is the resynchronization technology of the electron speed, and the other is technology of depressed collector that recovers the remaining energy of the electrons after the beam-wave interaction.

This chapter focuses on the technology of depressed collector. In the TWT, the efficiency can be increased to 30%~50% by using the speed resynchronization technology. It means that more than half of the electronic energy is still dissipated in the form of heat when the electrons are collected on the collector. Therefore, reducing this part of the energy loss will further improve the efficiency of the TWT. The potential of the collector determines the amount of energy consumed on the collector. If the potential of the collector is lower than the potential of the slow-wave structure, a deceleration field is formed between the slow-wave structure and the collector. The electrons are decelerated and give their own energy back to the power source before they hit the collector. Obviously, the lower the potential of the collector is, the stronger the deceleration field is, and the greater the electrons are decelerated, the more energy returned to the DC power is. This is the principle of using a depressed collector to improve the efficiency of a TWT. However, the collector potential cannot be reduced indefinitely, or it is not as low as possible. Because the speed of the electrons entering the collector region after interaction are scattered, and these electrons obey a certain velocity distribution, some of the electrons cannot overcome the deceleration field between the slow-wave circuit and the collector and these electrons return to the slow-wave circuit and disturb the normal operation of the TWT. Therefore, the multistage depressed collector [4–6] is used to improve the efficiency. The voltage of the collector is designed according to the electron groups with different speeds.

The relationship between collector voltage and current is shown in Figure 6.2. Consider the potential of the interaction zone as zero. When the potential gradually decreases from zero to V_1, the current does not change and remains at I_1. This means that all electrons in this interval can fly to the collector surface smoothly. In the case of no high-frequency field, the electron beam has the same speed when it reaches the entrance of the collector. When the potential of the collector is V_2, the collector receives all electrons, and the collector efficiency is 100%. If the collector potential continues to decrease, no electrons can be collected on the collector surface. At this time, the efficiency is at least zero. In practice, after the high-frequency field interacts with the electron beam, the speeds of the electrons in the electron beam are no longer uniform. The speed of some electrons is higher than the original electron speed, and the speed of other electrons may be lower than the original electron speed. Therefore, when the potential of collector drops to V_1, the slowest electron in the

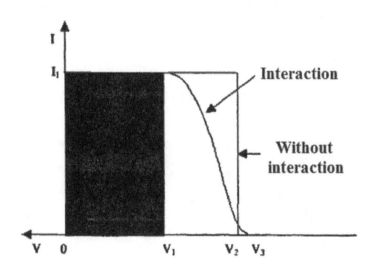

FIGURE 6.2 The relationship between voltage and current in collector.

electron beam just happens to reach the collector electrode. As the potential continues to decrease, more and more electrons cannot reach the collector surface, so the current of collector continues to decrease. When the potential of collector drops to V_3, there will be no electrons received on the collector plate, then the current of the collector will be zero. Assuming that the electrons do not reflow, the shaded area in Figure 6.2 indicates the maximum energy recovered by a single-stage depressed collector.

A well-designed collector usually uses several differently shaped electrodes at different potentials to selectively collect the electron beams that have been interacted with at low energy. This kind of collector is called multistage depressed collector. In practice, the spurs of the electron velocity are relatively large after coming from the interaction zone. A single-stage depressed collector can only recover very limited energy, and most of the interacted electronic energy is dissipated in the form of heat. In the case of very high recovery efficiency requirements, such as in a spacecraft communication system, the design using a single-stage depressed collector cannot meet the design requirements. Therefore, the electrons are needed to be classified according to different energy levels, and the corresponding electrodes are also set for electrons of different speeds to recover the electrons. Figure 6.3 shows a simple schematic of a four-stage depressed collector. Ideally, the energy recovered by the collector is the sum of the energy recovered on each electrode.

The diagram of the potential power recovery capability of a four-stage depressed collector for a completed electron beam is shown in Figure 6.4. The total recovered power is represented by the area of the shaded part. The formula is expressed as

$$P_{coll} = I_1 V_1 + I_2 \left(V_2 - V_1 \right) + I_3 \left(V_3 - V_2 \right) + I_4 \left(V_4 - V_3 \right) \tag{6.1}$$

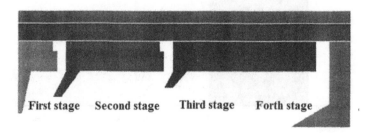

FIGURE 6.3 The schematic of four-stage depressed collector.

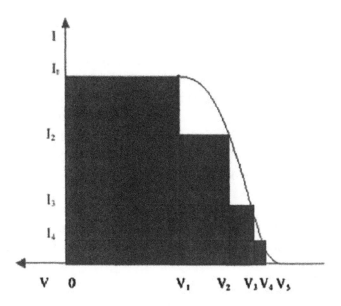

FIGURE 6.4   The total energy recovered by a four-stage depressed collector.

The efficiency of the TWT can be improved by increasing the number of electrodes, but the effect is not better with more stages. It is because more stages will increase the complexity of the manufacturing process of the microwave tube and power system. Therefore, a compromise must be considered to select the best number of stages. In general, the increment of the efficiency is already very small when the stage of the collector is more than 3 or 4. Thus, the number of stage of collector does not exceed 3 or 4 for most applications.

6.4 NUMERICAL ALGORITHM FOR COLLECTOR

6.4.1 The Basic Principle of the Algorithm

The collector is a component of the TWT electron optics system which includes electron gun, magnetic focusing system [7,8] and collector. It belongs to the field of intense electron optics. The distribution of the electric and magnetic fields of the system must be obtained in the analysis of the electron optics system. Therefore, the first problem in the research of the collector is to determine the distribution of the electric and magnetic fields of the collector. Due to the complexity of the electromagnetic field calculation, a computer program which is based on the basic equations of electron optics is needed to obtain accurate and reliable results. When the distribution of the electric field and magnetic field in the collector system is obtained, the movement trajectory of the electron in the collector can be known with the electron motion equation. Then, the laws of motion of the electron with electromagnetic field in the collector can also be obtained.

The basic equations of intense electron optics consist of three parts: Maxwell equation, the equation of electron motion in electromagnetic fields and continuity equation of electric current. They are the basis for studying the laws of motion in electron optics systems.

6.4.1.1 Maxwell Equation

In general, only static electromagnetic fields are studied in intense electron optics. The forms of Maxwell equation under static conditions are expressed as

$$\nabla \cdot \mathbf{D} = \rho \tag{6.2}$$

$$\nabla \times \mathbf{E} = 0 \tag{6.3}$$

$$\nabla \times \mathbf{H} = \mathbf{J} \tag{6.4}$$

$$\nabla \cdot \mathbf{B} = 0 \tag{6.5}$$

where $\mathbf{D} = \varepsilon_0 \mathbf{E}$ is electric displacement vector, $\mathbf{E}$ is electric intensity vector, H is magnetic intensity vector, $\mathbf{B} = \mu_0 \mathbf{H}$ is magnetic induction intensity vector, $\mathbf{J} = \sigma \mathbf{E}$ is current density vector, ρ is space charge density of free electron, σ is conductivity of dielectric, ε_0 is dielectric constant of vacuum and μ_0 is permeability of vacuum.

According to irrotationality of electrostatic field, electric scalar potential U can be expressed as

$$\mathbf{E} = -\nabla U \tag{6.6}$$

Then, equation $\nabla \cdot \mathbf{D} = \rho$ can be reduced to

$$\nabla^2 U = -\rho / \varepsilon \tag{6.7}$$

The equation mentioned above is a Poisson equation. The equation can be simplified to Laplace equation at $\rho = 0$.

$$\nabla^2 U = 0 \tag{6.8}$$

According to passivity of magnetostatic field in equation $\nabla \cdot \mathbf{B} = 0$, the magnetic vector potential A can be expressed as

$$\mathbf{B} = \nabla \times \mathbf{A} \tag{6.9}$$

Then, $\nabla \times \mathbf{H} = \mathbf{J}$ can be written as

$$\nabla \times \left(\frac{1}{\mu} \nabla \times \mathbf{A} \right) = \mathbf{J} \tag{6.10}$$

The expression of magnetic vector potential at a certain point P_0 for any magnetic system in the vacuum is written as

$$\mathbf{A} = \frac{\mu_0}{4\pi} \int_V \frac{1}{r} \mathbf{J}_T \, dV \tag{6.11}$$

where $\mathbf{J}_T$ is the whole current density vector which includes conduction current and displacement current, r is the distance from the point P_0 to volume element dV, μ_0 is the permeability of vacuum and V is the volume of the magnetic system.

6.4.1.2 Equation of Motion

The equation of motion of electrons in an electromagnetic field is expressed by

$$\frac{d}{dt}(m\mathbf{v}) = -e(\mathbf{E} + \mathbf{v} \times \mathbf{B}) \tag{6.12}$$

where t is the motion time, m is the electron mass, e is the electric charge of electron and $\mathbf{v}$ is the electron velocity.

6.4.1.3 Current Continuity Equation

The current density vector can be defined as

$$\mathbf{J} = \rho \mathbf{v} \tag{6.13}$$

The space charge is conserved. That is to say, the charge will not be generated or disappear out of nothing. Therefore, the current continuity equation is written as

$$\nabla \cdot \mathbf{J} = \nabla \cdot (\rho \mathbf{v}) = 0 \tag{6.14}$$

Combining Maxwell equation, current continuity equation, and electron motion equation, the basic equations for intense electron optics are expressed by

$$\nabla^2 U = -\rho / \varepsilon$$

$$\nabla \times \left(\frac{1}{\mu} \nabla \times \mathbf{A} \right) = \mathbf{J}$$

$$\frac{d}{dt}(m\mathbf{v}) = -e(\mathbf{E} + \mathbf{v} \times \mathbf{B}) \tag{6.15}$$

$$\nabla \cdot \mathbf{J} = 0$$

Solving the basic equations mentioned above, the motion state of the electrons in the collector can be obtained, and then the characteristics of the collector can also be analysed.

The essence of the energy recovered in the multistage depressed collector is that the electrons are decelerated in the electrostatic magnetic field, and the potential energy of the electrons is increased by reducing the kinetic energy. Therefore, it is necessary to satisfy the Poisson equation, the electron motion equation and the current continuity equation. They are self-consistent equations. At the beginning, when electrons enter the collector, the potential inside the collector only depends on the potential on the stage. As more electrons enter the collector, the potential inside the collector is affected by the space charge effect and then changed. At the same time, this change has affected the trajectory of the electrons. This change continues until the electric field reaches a steady state, that is to say the potential has no longer changed. Due to the complex shape of the boundary of the multistage depressed collector, it is impossible to analytically solve the system of electromagnetic field equations and the electron motion equation, so numerical calculation methods are needed.

The characteristic of the particle-in-cell method is to track the motion of macro particles by solving the complete Maxwell equations and Lorentz equations from the most basic electromagnetic field motion laws and the mechanical laws of particle motion. So, it can reflect the actual physical process more accurately. The particle-in-cell method is particularly suitable for problems that are not physically clear and difficult to analyse mathematically. It is also very adaptable for the cases with complex geometries, various boundary conditions and initial conditions. Therefore, it can be used not only for the study of basic theoretical issues in which the physical laws are not yet clear but also for the research and design of practical plasma devices with complex geometric shapes and structures. In general, the particle-in-cell method will be the most effective method for tracking the trajectory of electrons in the collector.

6.4.2 Secondary Electrons in the Collector

In order to improve the conversion efficiency, the TWT recovers energy from the interacted electrons through the collector. However, it will inevitably generate secondary electrons [9–12] when the electrodes of each collector collect electrons. The generation of secondary electrons is related to the energy, angle of the original electron incident on the electrode surface and the characteristics of the electrode surface. If secondary electrons hit a high-potential electrode and absorb energy, it will reduce the efficiency of the collector. If secondary electrons return to the high-frequency interaction zone, the heat dissipation power will be increased and noise power will generate in the high-frequency output window, even the entire tube will be burn out. An effective collector means that it can collect electrons with various speeds by the corresponding voltage electrode. It is to say that no electrons return to the interaction zone, and the speed of the electrons are zero when they reach the electrode surface, then fewer secondary electrons are generated. Therefore, it is positive for tube designers to design multistage depressed collectors to analyse quantitatively the effect of secondary electrons on efficiency of the collector under different electrode materials and voltages by using computer simulation technology, combining the secondary electron emission model theoretically.

The trajectories of secondary electrons are determined by the electric field distribution (including space charge effects) of the multistage depressed collector and the initial velocity of the secondary electrons. The original initial electrons hit on the surface of the electrode with different energies and angles, resulting in secondary electrons with different energies and angles at the point of impact. The distribution of energy and angle of secondary electrons is a function of primary electron energy and angle, electrode material and surface characteristics.

The main reasons for reducing of the efficiency of the multistage depressed collector are listed as follows.

a. Because the distribution of the electric field in the collector is not ideal, the electric field near the electrode surface cannot suppress the low-energy secondary electrons and even accelerate them to other electrodes. It can be avoided by using computer-aided design software. By observing the electric field distribution of typical secondary electrons and the working range of multistage depressed collectors, fine design can be made and secondary electrons are suppressed.

b. The angle of incidence of electrons at the bottom of the electrode with the lowest voltage is large.

c. Electrode surface or material with high secondary electron yield (SEY) is used. Therefore, electrode with lower SEY can be chosen.

6.5 SIMULATED EXAMPLES OF COLLECTOR

6.5.1 The Efficiency of the Collector

Figure 6.5 shows the various components of the power flow in a TWT. The definition of each variable is as follows. P_h is the heating power of the cathode filament, P_{RFin} is the input power of signal, P_0 is the power in the

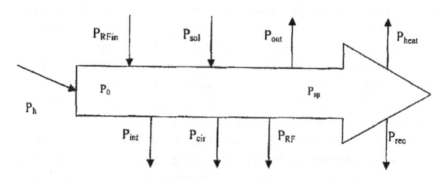

FIGURE 6.5   The power flow in a traveling wave tube.

electron beam generated by the electron gun, P_{int} is the power intercepted by the slow-wave system, P_{cir} is the loss power in circuit, P_{RF} is the loss power of radio frequency, P_{sol} is the power for focusing, P_{out} is the output power of radio frequency, P_{sp} is the power of waste electronics, P_{heat} is the power which is provided to the collector, and P_{rec} is the power which is recovered by the collector.

The supplied power includes the DC power which is supplied to the electron beam, focusing system and collector. The radio frequency input power is usually 30 dB lower than the output power, so it can be considered as negligible. When the electron beam is intercepted by the circuit or there is radio frequency loss, some power is lost. In addition, some radio frequency power generated by harmonic and intermodulation is also lost because it is not in the working frequency range. Finally, the electron beam (waste electron beam) leaving the high-frequency circuit enters the collector, after the high-frequency signal is output from the output coupler.

The relationship between power flow and efficiency can be derived according to the components of the power flow. So the efficiency of the whole tube is defined by the ratio of radio frequency output power to total input power.

$$\eta_{ov} = \frac{P_{out}}{P_{in}} = \frac{P_{out}}{P_{ot} - P_{rec}} = \frac{P_{out}}{(P_0 + P_h + P_{sol}) - P_{rec}} \qquad (6.16)$$

where P_{ot} is the power which is required to generate and maintain an electron beam. The power recovered by the collector can be expressed as $P_{rec} = \eta_{coll} \cdot P_{sp}$. Here, η_{coll} means the efficiency of the collector. The energy from the waste electron beam is the remaining energy which is from the electron gun after passing through the interaction zone. The expression of the energy is written as

$$P_{sp} = P_0 - P_{int} - P_{cir} - P_{RF} - P_{out} \qquad (6.17)$$

Thus,

$$P_{rec} = \eta_{coll} \cdot (P_0 - P_{int} - P_{cir} - P_{RF} - P_{out}) \qquad (6.18)$$

Here, two efficiencies are introduced. One is efficiency of the electron η_e, which varies from a few per cent to 50%. It refers to the efficiency of converting electron beam power to power at a specified frequency. Then, an expression is obtained.

$$P_{out} + P_{cir} = \eta_e P_0 \tag{6.19}$$

The other is the efficiency of circuit, which is usually between 75% and 90%. It is the efficiency with which radio frequency circuit transmits its power at a specified frequency to the output port. Thus,

$$P_{out} = \eta_{cir}\left(P_{out} + P_{cir}\right) = \eta_{cir}\eta_e P_0 \tag{6.20}$$

Then,

$$\eta_{ov} = \frac{\eta_e \cdot \eta_{cir}}{\dfrac{P_{ot}}{P_0} - \eta_{coll}\left[1 - \eta_e - \dfrac{P_{int} + P_{RF}}{P_0}\right]} \tag{6.21}$$

The formula mentioned above shows the relationship between efficiency and power. From which, it can be seen that the overall efficiency depends on efficiency of the collector, especially when the efficiency of the collector is high.

The efficiency of the collector is expressed by

$$\eta_{coll} = P_{rec}/P_{coll} \tag{6.22}$$

where P_{coll} is the power of collector and P_{rec} is the power which is recovered. The power which is recovered can be written as

$$P_{rec} = \sum_{n=1}^{m}|V_0 - V_{en}|I_{en} \tag{6.23}$$

where V_0 is the potential of the slow-wave system, V_{en} is the potentials of each electrode of the collector, I_{en} is the current of each electrode of the collector, and m is the number of the electrode in the collector.

The formula of calculating the power of the collector is written as

$$P_{coll} = V_0 I_0 - P_{body} - P_{RF} \tag{6.24}$$

where P_{RF} is the output power of radio frequency.

6.5.2 Simulated Example

In this chapter, a model of the collector is shown in Figure 6.6. The efficiency of the collector of the TWT is calculated with electromagnetic particle software. In order to obtain higher efficiency and reduce the impact of the secondary electron emission on the collector during the working process of the collector, the multistage depressed collector is used. From Figure 6.6, it can be seen that a four-stage depressed collector is simulated.

During simulation, the characteristics of secondary electron emission of the material for the collector should also be determined, in addition to paying attention to the state (including speed and position) of all electrons which is at entrance of the collector. For most of TWTs, the collectors are made of oxygen-free copper. Therefore, the efficiency of collector of a TWT with oxygen-free copper is analysed.

As can be seen from the previous chapters, the SEY [13] of the material can be changed by modifying the surface morphology through surface modification. Based on the conclusions of this study, a chemical etching process is used to construct a regular array structure on the surface of oxygen-free copper samples. Then, the characteristic of the secondary electron emission is studied.

The lithography process is a mature planar process method for realizing micro–nano structures in the field of semiconductor. The typical process flow chart is shown in Figure 6.7. First, the photoresist with a certain thickness is spin-coated on the surface of the sample. Then, the selected photoresist is exposed by using ultraviolet light through the mask. It is noticed that the light-transmitting area on the mask is a pre-designed pattern. Second, the exposed photoresist is etched away with the developer, so as to expose the surface of the substrate for selective chemical etching of the sample. Lastly, a pattern which is consistent with the mask designed

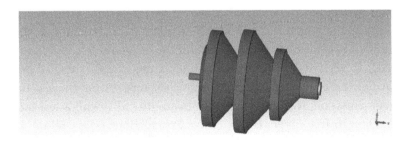

FIGURE 6.6   The model of collector.

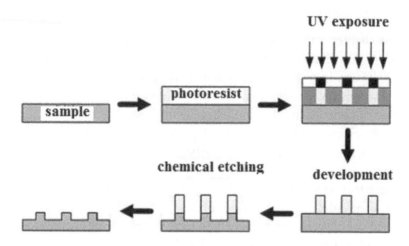

FIGURE 6.7   The process of flow chart.

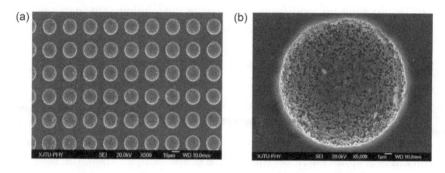

FIGURE 6.8   The sample with regular array structure. (a) Micropore array, (b) Single micropore.

is obtained on the sample surface by removing the remaining photoresist. The parameters of the structure are that the gap between the microstructures is 7 μm, the diameter of the microcylinder is 15 μm and the height is 5 μm.

By using the process flow above, the regular array structure, as shown in Figure 6.8, is formed on the surface of the oxygen-free copper sample. In order to verify the effect of the regular array structure etched on the surface of the oxygen-free copper sample on suppressing the secondary electron emission, it is necessary to measure the SEY with a test instrument. Oxygen-free copper is susceptible to oxidation and is contaminated on the surface. In order to obtain the SEY of oxygen-free copper under ideal conditions, the acetone solution is used to clean the oxygen-free

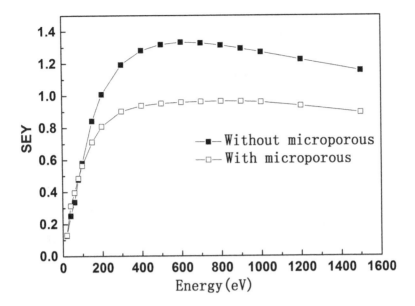

FIGURE 6.9   SEY of the oxygen-free copper samples.

copper sample with ultrasonic before testing. For comparison, SEYs of the oxygen-free copper samples etched and not etched are tested with the test platform mentioned in the previous section. The results are shown in Figure 6.9.

Figure 6.9 shows the measured curves of SEY of oxygen-free copper samples. From the test results, it can be seen that the SEY value of the oxygen-free copper sample etched is lower than that of the oxygen-free copper sample without etched. The maximum value decreases from 1.33 to 0.96 by etching. The overall SEY value of the oxygen-free copper sample etched is less than 1, which means that it will not generate secondary electrons when electrons hit the surface of the sample. The test results also show that the etching process can reduce the SEY of oxygen-free copper indeed, and the effect of suppression is obvious.

Based on the test results in Figure 6.9 and the electromagnetic particle simulation method, the collector model in Figure 6.6 is analysed. The voltages of the four electrodes of the collector are shown in Table 6.1. During simulation, it is assumed that the working voltage of the TWT is 280 kV and the working current of the electron beam is 11 A. Meanwhile, it is assumed that the position and velocity of the electrons at the entrance of the collector satisfy a uniform distribution.

TABLE 6.1　The Voltages of the
Four Electrodes of the Collector

Stage	Voltage (V)
First stage V_1	2.50e5
Second stage V_2	1.50e5
Third stage V_3	2.00e4
Fourth stage V_4	−5.00e4

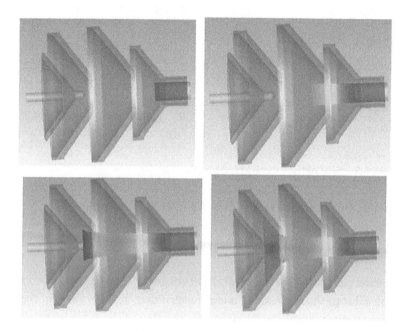

FIGURE 6.10　The processes of collecting electrons in collector.

The motion of the electrons in the collector model during simulation is shown in Figure 6.10. As can be seen from the figure, secondary electrons are inevitably generated during the process of collecting electrons. The efficiency of this collector [14] can be calculated according to the given conditions. By simulating, it is found that the efficiency of the collector with the surface-treated SEY value of the microporous sample is 80.1%, and the efficiency of the collector with the SEY value of the sample without micropores is 82.55%. The calculation results show that the efficiency of the collector with the SEY value of the etched surface is significantly improved.

6.6 SUMMARY

In this chapter, the working principle of collector in the TWT is introduced, and the importance of secondary electron emission in the collector is discussed. Then, the application of the electromagnetic particle simulation method in the collector of TWT is described. Finally, the efficiency of the collector is calculated for a specific collector. The results show that the SEY has a very important effect on the collector of the TWT. The efficiency of the collector can be improved by reducing the coefficient of SEY.

REFERENCES

1. Liao FJ. *Vacuum Electronic Technology-The Heart of Information Weaponry.* Beijing: National Defense Industry Press, 2008, 1–26.
2. Wang WX. *Microwave Engineering Technology.* Chengdu: National Defense Industry Press, 2014, 333–441.
3. Bai CJ. *Study of Beam Wave Interaction Basic Theory and CAD Technique for Coupled Cavity Traveling Wave Tube.* Chengdu: University of Electronic Science and Technology of China, 2013.
4. Gao XH. *Study on CAD Technique of Multistage Depressed Collector for Traveling Wave Tube.* Chengdu: University of Electronic Science and Technology of China, 2004.
5. Cai ZY. *Simulation and Design of Multistage Depressed Collector for Traveling Wave Tube.* Hefei: Hefei University of Technology, 2006.
6. Xu X, Yang J, Lv GQ, et al. Design and simulation of multistage depressed collector for space traveling wave tube. *Vacuum Electronics,* 2011, 01:1–8.
7. Yu YH. *Research on Design and Characteristics of Electron Optical System in Plasma Traveling Wave Tube.* Chengdu: University of Electronic Science and Technology of China, 2006.
8. Jin YB. *Electron Optical Emission Model and Numerical Simulation of Traveling Wave Tube.* Chengdu: University of Electronic Science and Technology of China, 2007.
9. He Q, Kou JY, Sun Y, et al. Study on secondary electron in multistage depressed collector. *Vacuum Electronics,* 2004, 01:25–28.
10. Ma YL. *The Research on Auxiliary Design and Secondary Electron of Multistage Depressed Collector for Traveling Wave Tube.* Chengdu: University of Electronic Science and Technology of China, 2006.
11. Ding MQ, Huang MG, Feng JJ, et al. Secondary electron emission suppression of multistage depressed collector in space traveling tube by Mo Ion Deposition. *Chinese Journal of Vacuum Science and Technology,* 2009, 29(3):247–250.
12. Huang T, Yang ZH, Jin YB, et al. The emission model of secondary electron in multistage depressed collector CAD. *Journal of Electron Information Technology,* 2008, 30(5):1247–1250.

13. Ye M, He YN, Wang R, et al. Suppression of secondary electron emission by micro-trapping structure surface. Acta Phys. Sin, 2014, 63(14):147901.
14. Bai CJ, Cui WZ, Ye M, et al. To improve the efficiency of the collector of space traveling wave tube by reducing secondary electron yield with micromachining technology. *Chinese Space Science and Technology*, 2017, 37(2):36–42.